Vol. 28. **The Analytical Chemistry of Nitrogen and Its Compounds** (*in two parts*). Edited by C. A. Streuli and Philip R. Averell

Vol. 29. **The Analytical Chemistry of Sulfur and Its Compounds** (*in three parts*). By J. H. Karchmer

Vol. 30. **Ultramicro Elemental Analysis.** By Günther Tölg

Vol. 31. **Photometric Organic Analysis** (*in two parts*). By Eugene Sawicki

Vol. 32. **Determination of Organic Compounds: Methods and Procedures.** By Frederick T. Weiss

Vol. 33. **Masking and Demasking of Chemical Reactions.** By D. D. Perrin

Vol. 34. **Neutron Activation Analysis.** By D. De Soete, R. Gijbels, and J. Hoste

Vol. 35. **Laser Raman Spectroscopy.** By Marvin C. Tobin

Vol. 36. **Emission Spectrochemical Analysis.** By Morris Slavin

Vol. 37. **Analytical Chemistry of Phosphorus Compounds.** Edited by M. Halmann

Vol. 38. **Luminescence Spectrometry in Analytical Chemistry.** By J. D. Winefordner, S. G. Schulman and T. C. O'Haver

Vol. 39. **Activation Analysis with Neutron Generators.** By Sam S. Nargolwalla and Edwin P. Przybylowicz

Vol. 40. **Determination of Gaseous Elements in Metals.** Edited by Lynn L. Lewis, Laben M. Melnick, and Ben D. Holt

Vol. 41. **Analysis of Silicones.** Edited by A. Lee Smith

Vol. 42. **Foundations of Ultracentrifugal Analysis.** By H. Fujita

Vol. 43. **Chemical Infrared Fourier Transform Spectroscopy.** By Peter R. Griffiths

Vol. 44. **Microscale Manipulations in Chemistry.** By T. S. Ma and V. Horak

Vol. 45. **Thermometric Titrations.** By J. Barthel

Vol. 46. **Trace Analysis: Spectroscopic Methods for Elements.** Edited by J. D. Winefordner

Vol. 47. **Contamination Control in Trace Element Analysis.** By Morris Zief and James W. Mitchell

Vol. 48. **Analytical Applications of NMR.** By D. E. Leyden and R. H. Cox

Vol. 49. **Measurement of Dissolved Oxygen.** By Michael L. Hitchman

Vol. 50. **Analytical Laser Spectroscopy.** Edited by Nicolo Omenetto

Vol. 51. **Trace Element Analysis of Geological Materials.** By Roger D. Reeves and Robert R. Brooks

Vol. 52. **Chemical Analysis by Microwave Rotational Spectroscopy.** By Ravi Varma and Lawrence W. Hrubesh

Vol. 53. **Information Theory As Applied to Chemical Analysis.** By Karel Eckschlager and Vladimir Štěpánek

Vol. 54. **Applied Infrared Spectroscopy: Fundamentals, Techniques, and Analytical Problem-solving.** By A. Lee Smith

Vol. 55. **Archaeological Chemistry.** By Zvi Goffer

Vol. 56. **Immobilized Enzymes in Analytical and Clinical Chemistry.** By P. W. Carr and L. D. Bowers

Vol. 57. **Photoacoustics and Photoacoustic Spectroscopy.** By Allan Rosencwaig

Vol. 58. **Analysis of Pesticide Residues.** Edited by H. Anson Moye

Vol. 59. **Affinity Chromatography.** By William H. Scouten

Vol. 60. **Quality Control in Analytical Chemistry.** By G. Kateman and F. W. Pijpers

Vol. 61. **Direct Characterization of Fineparticles.** By Brian H. Kaye

Vol. 62. **Flow Injection Analysis.** By J. Ruzicka and E. H. Hansen

Vol. 63. **Applied Electron Spectroscopy for Chemical Analysis.** Edited by Hassan Windawi and Floyd Ho

(*continued on back*)

# Kinetic Aspects of Analytical Chemistry

# CHEMICAL ANALYSIS

## A SERIES OF MONOGRAPHS ON
## ANALYTICAL CHEMISTRY AND ITS APPLICATIONS

**VOLUME 96**

**A WILEY-INTERSCIENCE PUBLICATION**

### JOHN WILEY & SONS

**New York / Chichester / Brisbane / Toronto / Singapore**

# Kinetic Aspects of Analytical Chemistry

**Horacio A. Mottola**

Department of Chemistry
Oklahoma State University

**A WILEY-INTERSCIENCE PUBLICATION**

**JOHN WILEY & SONS**

**New York** / **Chichester** / **Brisbane** / **Toronto** / **Singapore**

Copyright © 1988 by John Wiley & Sons, Inc.

All rights reserved. Published simultaneously in Canada.

Reproduction or translation of any part of this work
beyond that permitted by Section 107 or 108 of the
1976 United States Copyright Act without the permission
of the copyright owner is unlawful. Requests for
permission or further information should be addressed to
the Permissions Department, John Wiley & Sons, Inc.

*Library of Congress Cataloging in Publication Data:*
Mottola, Horacio A.
   Kinetic aspects of analytical chemistry / Horacio A. Mottola.
   p. cm.—(Chemical analysis; v. 96)
   "A Wiley-Interscience publication."
   Bibliography: p.
   Includes index.
   ISBN 0–471–83676–1
   1. Chemistry, Analytic.   2. Chemical reaction, Rate of.
   I. Title. II. Series.
QD75.4.K54M68 1988   87-26344 CIP
543—dc 19

Printed in the United States of America
10 9 8 7 6 5 4 3 2 1

This book is dedicated to the memory of the late Ernest B. Sandell (University of Minnesota), who unexpectedly died on March 10, 1984 in Houston, when returning from a trip to Mexico. He played a key role in promoting the use of catalytic methods of determination, and guided me in my first steps into research in analytical chemistry. His retiring personality prevented him from gaining deserved recognition.

# PREFACE

Kinetics has been well recognized in most traditional subdisciplines of chemistry (physical, organic, and inorganic) but was neglected by analytical chemists, mostly because of the complete domination of teaching in that area by the equilibrium viewpoint of chemical reactions. During the years in which gravimetric and titrimetric procedures represented the bulk of analytical methodology available to the practicing chemist, such a viewpoint undoubtedly was largely justified. However, developments in electroanalytical chemistry (e.g., polarography) and analytical spectroscopy (ultraviolet and visible absorption), for instance, provided the first sparks of change. From there, concepts of reaction kinetics and the possibility of obtaining useful information with measurements under dynamic conditions spread into analytical chemistry in roughly the late 1930s and early 1940s. The accelerating impact of operational amplifiers on the development of modular analytical instrumentation with capabilities of processing temporal signals reliably, in the late 1950s and early 1960s, further opened avenues for kinetic thinking and measurement in analytical chemistry. The richness of kinetics goes beyond its application in determinations; after all, any physical or chemical process in contemporary analytical chemistry has temporal connotations.

This book discusses in an integrated form those aspects of kinetics that have become part of today's analytical chemistry. It is an outgrowth of lectures given in the second half of a one-semester course for first-year graduate students entitled "Equilibrium and Kinetics in Analytical Chemistry." This course is offered at Oklahoma State University as part of its core course requirement for students majoring in analytical chemistry. The overview presented by Chapter 1 is aimed at defining the scope of the text, while succeeding chapters delve into each aspect of relevance, with focus on the analytical side of the fundamentals and applications rather than on kinetics as treated in classical books on physical chemistry, biochemistry, and the like.

This book could not have been written without the work of those cited in the reference sections of each chapter. I am particularly grateful to J. D. Winefordner for his invitation to write the book, for encouragement during its preparation, and for reading Chapter 8; to Otis C. Dermer for reading, chapter by chapter, the entire manuscript; and to H. L. Pardue for reading Chapter 10. I first toyed with kinetic methods as a postdoctoral research associate in Henry Freiser's research group at the University of Arizona, and the opportunity for this is gratefully acknowledged here.

HORACIO A. MOTTOLA

*Stillwater, Oklahoma*
*January 1988*

# CONTENTS

CHAPTER 1    AN OVERVIEW OF KINETICS IN
ANALYTICAL CHEMISTRY    1

1.1.   Historical View of the Evolution of Kinetic-Based Methods of Determination    1

1.2.   Chemical Reactions and Physical Processes in Analytical Chemistry    14

    1.2.1.   Chemical Reactions    15

    1.2.2.   Physical Processes    19

    References    21

CHAPTER 2    CATALYTIC METHODS: NONENZYMATIC METHODS    24

2.1.   Mathematical Basis of Catalytic Methods Based on Primary Catalytic Effects    25

2.2.   Practical Treatment of Data in Catalytic Determinations    29

2.3.   Chemical Reactions for Catalytic Determinations    36

2.4.   Limits of Detection, Sensitivity, Selectivity, and Optimization in Catalytic Determinations    40

    2.4.1.   Limits of Detection and Sensitivity    40

    2.4.2.   Selectivity and Specificity    42

    2.4.3.   Optimization of Catalytic Methods    44

2.5.   Some Applications of Nonenzymatic Catalytic Determinations    45

    2.5.1.   The Sandell–Kolthoff Reaction    45

    2.5.2.   Miscellaneous Applications of Catalyzed Reactions    48

    References    50

CHAPTER 3    CATALYTIC METHODS: HOMOGENEOUS ENZYMATIC METHODS    53

3.1.   Enzymes as Catalysts    53

3.2.   Kinetics of Enzyme-Catalyzed Reactions    54

ix

3.3.   The Michaelis–Menten Constant                      57
3.4.   Enzyme Activity and Enzyme Concentration           58
3.5.   Analytical Applications of Enzyme-Catalyzed
       Reactions                                          62
       3.5.1.   Clinical Determinations of Enzyme
                Activities                                62
       3.5.2.   Determination of Substrates by Means
                of Enzyme-Catalyzed Reactions             63
3.6.   Chemical Approaches to the Monitoring of
       Enzyme-Catalyzed Reactions                         63
       References                                         70

CHAPTER 4   METHODS OF DETERMINATION BASED ON
            MODIFIED CATALYZED REACTION RATES
            IN SOLUTION                                   72

4.1.   Modification of Metal-Ion-Catalyzed Systems        72
       4.1.1.   Applications of Inhibition                73
       4.1.2.   Applications of Activation                76
       4.1.3.   Applications of Promoting Effects         78
       4.1.4.   Oscillating Reactions                     80
4.2.   Modification of Enzyme-Catalyzed Systems           81
       4.2.1.   Activation                                81
       4.2.2.   Inhibition                                83
       References                                         85

CHAPTER 5   ANALYTICAL APPLICATIONS OF
            HETEROGENEOUS CATALYSIS                       88

5.1.   Heterogeneous Catalyzed Electrode Reactions        89
       5.1.1.   Coupled Chemical Reactions                89
       5.1.2.   Analytical Applications of Voltam-
                metric Catalytic Currents                 92
5.2.   Heterogeneous Catalytic Determinations
       Based on Immobilized Enzymes                       93
       5.2.1.   Enzyme Regeneration in Homogeneous
                Systems                                   93
       5.2.2.   Enzyme Regeneration by Use of Immo-
                bilized Enzyme Preparations               99
       References                                        108

CHAPTER 6    **RATE DETERMINATIONS USING UNCATALYZED REACTIONS**    **111**

6.1.   Rate Considerations for the Determination of a Single Species in Uncatalyzed Processes    111

6.1.1.   Pseudo-Zero-Order Conditions (Initial Rate Method)    111

6.1.2.   First-Order and Pseudo-First-Order Conditions    112

6.2.   Special Cases for the Determination of a Single Species in a Mixture by Use of an Uncatalyzed Reaction    113

6.3.   Selected Analytical Applications of Rates of Uncatalyzed Reactions    115

6.3.1.   Determination of Inorganic Species    115

6.3.2.   Determination of Organic Species    115

References    120

CHAPTER 7    **DIFFERENTIAL REACTION RATE METHODS**    **122**

7.1.   Methods Based on First-Order or Pseudo-First-Order Kinetics (Region I of Figure 7.1)    124

7.1.1.   Logarithmic Extrapolation Method    124

7.1.2.   Single-Point Method    126

7.1.3.   Tangent Method    127

7.1.4.   Method of Proportional Equations    128

7.2.   Methods Based on Second-Order Kinetics (Regions III, IV, and V of Figure 7.1)    133

7.2.1.   Second-Order Logarithmic Extrapolation    133

7.2.2.   Second-Order Linear Extrapolation (Region IV), Second-Order Single-Point (Region IV), and Second-Order Graphical Differential (Tangents) and Integral (Regions III, IV, and V) Methods    134

7.3.   Pseudo-First-Order Methods with Respect to a Common Reagent (Region VII of Figure 7.1)    135

7.3.1.   Single-Point Method of Roberts and Regan    135

7.3.2. Method of Proportional Equations      136

7.4. Pseudo-Zero-Order Methods (Regions I
through VII of Figure 7.1)                 136

7.4.1. Single-Point Method                136

7.4.2. Method of Proportional Equations    137

7.5. Critical Evaluation of Selected Differential
Reaction Rate Methods                     137

7.6. Application of Continuous-Flow
Sample–Reagent(s) Processing in Implementing Differential Reaction Rate Procedures  137

7.7. Some Miscellaneous Approaches to
Differential Determinations               140

References                                146

CHAPTER 8    KINETIC METHODS BASED ON DETECTION
OF LIGHT EMISSION: FLUORESCENCE,
PHOSPHORESCENCE, CHEMILUMINE-
SCENCE, AND BIOLUMINESCENCE              149

8.1. Fluorescence and Phosphorescence       149

8.1.1. Fluorescence                       149

8.1.2. Phosphorescence                    153

8.2. Chemiluminescence and Bioluminescence  158

References                                167

CHAPTER 9    INSTRUMENTATION                        170

9.1. Means of Mixing Reactants              171

9.1.1. Mixing by Manual and Magnetic
Stirring                            171

9.1.2. Stopped-Flow Mixing                172

9.1.3. Continuous-Flow Mixing             178

9.1.4. Mixing for Kinetic-Based Titrimetry
and Stat Procedures                 184

9.1.5. Centrifugal Mixing                 187

9.2. Detection Approaches                   189

9.2.1. Absorptiometric Detection          189

9.2.2. Detection by Means of Fluorescence,

Chemiluminescence, and Biolumine-
scence                                          193
9.2.3.  Electrochemical Detection               194
9.2.4.  Other Methods                           197
9.3.  Ancillary Electronic Units                198
9.3.1.  Circuit for Automatic Measurement of
Slopes of Rate Curves                           198
9.3.2.  A Fixed-Time Digital Counting System    199
9.3.3.  All-Electronic Reciprocal Time
Computers                                       201
9.3.4.  Systems for Differential Rate
Measurements                                    201
9.3.5.  Analog Systems for Catalytic End-Point
Detection and Miscellaneous
Switching Systems                               203
9.4.  Computers                                 207
References                                      207

CHAPTER 10    ERROR ANALYSIS IN KINETIC-BASED
DETERMINATIONS                                  212

10.1.  Minimization of Systematic and Random
Fluctuations in Rate Coefficients               212
10.1.1.  Single-Rate Measurement Approach       212
10.1.2.  Two-Rate Measurement Approach          216
10.1.3.  Multirate Measurements: Multipoint
Linear Regression Treatment with
Predictive Equilibrium Signal Values            218
10.1.4.  Optimization of Irreversible, Coupled
First-Order Reactions                           222
10.2.  Regression Computations in Differential
Reaction Rate Methods                           223
10.2.1.  Least-Squares Regression               226
10.2.2.  Nonlinear Regression                   226
10.3.  Comparative Error Studies                227
10.4.  Kalman Filtering                         230
References                                      233

**CHAPTER 11 KINETIC COMPONENTS IN SEVERAL
ANALYTICAL TECHNIQUES OR STEPS
IN ANALYSIS                                    236**

11.1. Diffusion in Analytical Chemistry          238

11.2. Kinetics in Chromatography                 240

11.3. Kinetics in Electrochemical Processes      242

11.4. Kinetics in Absorption/Emission
Spectroscopy                               244

    11.4.1. Series Processes in Nonflame Atomic
Absorption Spectroscopy          245

    11.4.2. Other Kinetic Considerations of
Interest in Flameless Atomization   246

    11.4.3. Kinetics in Analytical Flame
Spectroscopy                     248

11.5. Kinetics in Continuous-Flow
Sample–Reagent(s) Processing               248

    11.5.1. Kinetics in Air-Segmented
Sample–Reagent(s) Processing Systems   249

    11.5.2. Kinetics in Single-Channel Unsegmen-
ted Sample–Reagent(s) Processing
Systems                          252

    11.5.3. Kinetics in Unsegmented Continuous-
Flow Systems Using Mixing
Chambers                         254

11.6. Other Kinetic Aspects of Relevance
in Analytical Chemistry                    258

References                                 259

**EPILOGUE                                       263**

**INDEX                                          265**

# Kinetic Aspects of Analytical Chemistry

CHAPTER

1

# AN OVERVIEW OF KINETICS IN
# ANALYTICAL CHEMISTRY

## 1.1. HISTORICAL VIEW OF THE EVOLUTION OF KINETIC-BASED METHODS OF DETERMINATION

The analytical chemist's perception of kinetics as a useful tool is marred by a provincial attitude that still prevails in some sections of analytical chemistry. This attitude considers that what is truly analytical is only concerned with qualitative detection or quantitative determination. No doubt this restricted concept is reflected in the indiscriminate use of the word *analysis* by many analytical chemists as well as the neglect of formal training in important analytical steps such as sampling. It is not surprising then that the first extensive historical account of analytical chemistry (1) records "catalytic analysis" as the only entry of kinetic nature and as a subdivision of microanalysis. Catalytic determinations undoubtedly make lower relative amounts (percentages) of constituents available for determination by rather conventional techniques; hence the relatively early impact catalysis had in analytical chemistry.

Catalytic determinations, however, are applicable also to macrosamples [a sample of weight greater than 0.1 g (2)], and their classification as part of microanalysis is questionable. In any event, catalytic determinations broke the pattern prevailing in the years when gravimetric and titrimetric procedures dominated the practice of analytical chemistry. In those years, kinetics was considered mainly in the context of undesirable effects (e.g., the sluggishness of attainment of some end points in redox titrimetry). The presence of kinetic concepts was very limited in those years, although an outstanding example can be singled out: it was recognized that the rate of crystal growth greatly influences the ease of filtration and the purity of precipitates. The actual kinetic formulation and a great deal of clarification of the mechanism of the precipitate formation and growth had to wait, however, until the 1950s and early 1960s (3). Precipitation from homogeneous solution, an ingenious

From a plenary lecture presented at the First International Symposium on Kinetics in Analytical Chemistry, Córdoba, Spain, September 27–30, 1983. Reproduced with permission from *Química Analítica*, official publication of the Spanish Society of Analytical Chemistry.

application of chemical kinetics, was reborn in the 1930s (4) and well documented in the 1950s (5). Its first use seems to date back to about a century earlier (Chancel in 1858, Ref. 1, p. 184), apparently following a pattern rather than constituting an exception: most discoveries or ideas appear to need an induction period before they are noticed and applied by the practitioner.

Catalytic determinations did not escape the induction period pattern. As early as 1876, Guyard (also known by the alias Hugo Hamm) discussed the catalytic effect of vanadium on the chlorate ion oxidation of aniline to yield aniline black, and he proposed the use of this reaction to detect the presence of vanadium (6). It took a decade, however, until Witz and Osmond described a semiquantitative way of estimating the amount of vanadium by assuming that the amount of aniline black formed after a given time is proportional to the amount of vanadium present in the sample (7). The report of their findings and of the analytical method was made by Osmond after Witz's death. They dipped oxycellulose threads into vanadium-containing standard solutions and subsequently into an aniline–chlorate mixture and reported a practical limit of detection of 5 ng of vanadium. This can be considered the first effort to quantify a catalytic procedure, and it singles out the fixed-time approach as the first one used in catalytic measurements.

As already noted, the familiar equilibrium-based determinations dominated analytical chemistry at the time of Osmond and Witz, and about 50 years elapsed before Sandell and Kolthoff reported on the catalytic action of iodide on the cerium(IV) oxidation of arsenic(III) and proposed its use for the estimation of iodide (8). A few years later they provided specific conditions for the catalytic determination of iodide by a variable-time procedure [at that time called chronometric determination in recognition of the time measurement (9)]. This was a landmark contribution to adding kinetics to the arsenal of analytical chemistry; it appeared at a fruitful time since almost simultaneously in Hungary, Szebelledy and Ajtai reported on the vanadium catalysis of the chlorate oxidation of $p$-phenetidine and the activating action of hydroxytartrate ions (10). Later Szebelledy and Ajtai suggested use of the bromate oxidation of phenetidine in the presence of 1,2-benzenediol (pyrocatechol) for the determination of as little as 0.6 ng of vanadium (11).

A large variety of catalytic determinations have been described since these early reports; most of them found a place in the first monograph on kinetic methods of determinations authored by Yatsimirskii and published in 1966 (12). This monograph, a landmark in itself and reflecting an overwhelming dedication of Russian analytical chemists to catalytic determinations, is devoted almost entirely to catalytic methods.

The proliferation of catalytic methods has produced about 50 papers per year in recent years (13), more than 10 times as many applications as the total reported for any other single approach. Figure 1.1 illustrates the growth of

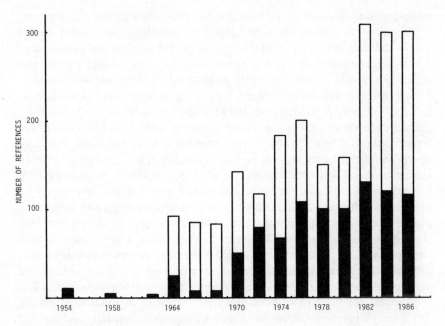

**Figure 1.1** Trends in the coverage of kinetic methods and miscellaneous aspects of kinetics in analytical chemistry as reflected in the biennial reviews in *Analytical Chemistry*. The solid portions of the bars indicate the number of references for catalytic methods in the given year.

catalytic determinations through the years, with information derived from the originally annual (now biennial) fundamental reviews in the journal *Analytical Chemistry*. These reviews have been chosen because they assure a certain degree of continuity. Although the first such reviews appeared in January 1949, no mention of catalytic determinations occurred until 1954, in West's review on inorganic chemistry (14). Specific reviews on kinetic aspects of analytical chemistry did not appear until 1964; since then kinetic methods and some other kinetic items of analytical interest have been regularly covered in these biennial reviews. The one published in 1970 also covers enzymatic determinations, but they are not taken into consideration in Figure 1.1, in order to make consistent comparison with the other reviews.

The reviews used to construct Figure 1.1 also reflect interesting trends in the evolution of the topics covered. Indirect determinations of organic chemical species (by a variety of inhibitory approaches) and stopped-flow mixing are noticeable in the reviews of the 1970s, and the increased role of electronic instrumentation and computers starts to be seen in the 1974 review. The 1984 review clearly points to the increased role that continuous-flow sample–reagent(s) processing is having in the analytical chemistry of solutions. Interest in continuous-flow systems occurs not only because of their inherent kinetic nature but because of their utility in providing a uniquely convenient way to automate kinetic-based determinations. The impact of the un-segmented version popularized by the name "flow injection analysis" was clear during the First International Symposium on Kinetics in Analytical Chemistry, and this is singled out in the 1984 review.

Catalytic determinations constitute the most widely used of the kinetic-based methods. Their dominance is more evident when enzymatic methods, which are also catalytic in nature, are considered. The other major area in kinetic methods of determination is differential reaction rate methods. In contrast with catalytic methods, differential methods are not aimed at determining ultra-low concentrations of materials in solution, and most of the reactions employed are uncatalyzed. The appearance of differential rate methodology can be traced to the mid-1940s, when a new dimension in kinetic-based determinations was added: the manipulation of kinetic expressions to extract concentration information on at least two species in the same sample by graphical methods and without resort to prior separation. The single-point approach of Kolthoff and Lee (15) greatly stimulated interest in this direction, demonstrating that kinetics, through differential reaction rate methods, offered means for the determination of medium and major components in a mixture without need of prior separation. Kolthoff and Lee's paper (15) describes a procedure for the determination of the external double bonds in synthetic rubbers, and the method is explored further in a subsequent paper by the same authors (16). It is based on the differences in rates of reaction of peroxybenzoic acid with internal and external double bonds according to the general scheme

The amount of peroxybenzoic acid that has not reacted to produce the epoxide plus benzoic acid is determined by a simple iodometric titration. Three methods of interpreting the results are discussed in the paper. The first is a graphical (not logarithmic) extrapolation, that is considered inaccurate. The second method, identified as the approximation method, is more accurate but more involved. The third method is the one recognized today as the single-point method and uses calibration curves (constructed following theoretical kinetic considerations) in which the percentage of double bonds reacted after 3 hr is plotted versus composition of known mixtures. It is of interest to note that this analytical procedure appeared in a journal not recognized as an analytical journal and that the procedure resulted from research sponsored by the Office of Rubber Reserve in connection with a government program on synthetic rubber.

The organic chemistry component in Kolthoff and Lee's papers is dominant. Publication of these papers documenting differential rate procedures preceded by a couple of years the appearance of the first book dealing with quantitative organic analysis based on functional group behavior (17). Organic chemical analysis was making its presence felt strongly, and so were differential reaction rate concepts.

The further development of differential reaction rate methods parallels the increasing acceptance of organic chemical analysis as part of analytical chemistry; this is well documented in the second monograph on kinetics in analytical chemistry (18). A whole chapter of this monograph is dedicated to the determination of organic mixtures by use of differential reaction rate procedures.

The historic paper by Kolthoff and Lee (15) also contains a statement of special interest:

A general method for the analytical interpretation of the experimental data obtained when a reagent reacts with a mixture of two compounds at considerably different rates has not been described in the literature.

As early as 1921, however, Clarens, by applying the laws of chemical kinetics to quantitative analysis (19), proposed a determination of tannins in general and in wine in particular; he also characterized three tannin fractions through analysis of the reaction curve obtained by plotting manometric readings (reaction with oxygen) as a function of time. The chemistry of tannins is recognized as complex and nonuniform; hence the revelation that fractions react with oxygen at different rates is not surprising. What is of special interest is that Clarens used the kinetic expression for first-order reactions to calculate the curve expected if a single tannin fraction were responsible for the response curve. Clarens then utilized extrapolation to time zero to estimate the corresponding fraction by use of the theoretically extrapolated curve (Figure 1.2).

Difficult as it is to establish historical priorities, Clarens's paper appears to describe one of the earliest uses of differential reaction rate methods and the use of integrated chemical rate expressions for the in situ determination of more than one species in solution. Kolthoff and Lee's work, however, did more to trigger the attention of other chemists; it initiated a period that extended

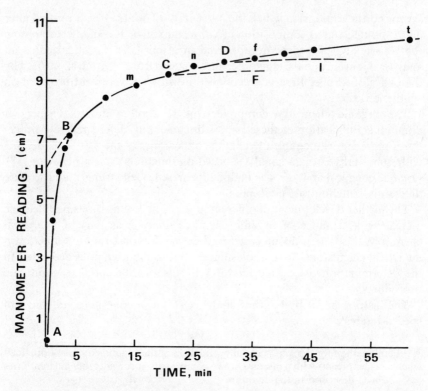

**Figure 1.2.** Plot of manometric reading versus time for the reaction of tannins with oxygen under good stirring. Retraced for the original (Ref. 19). The portion HBmF was calculated by use of the kinetic expression $k = (1/t)\log[a/(a-X)]$, where $X =$ quantity oxidized at time $t$, and $a =$ original quantity to be oxidized. Portion AB is attributed to the predominance of a fast-reacting tannin, and portion BC to a fraction reacting at intermediate rate. A similar consideration is offered for the extension CDI. $H$ is the expected intercept if all tannin present were the intermediate-reacting type characterized by the rate coefficient $k$.

through the 1950s and 1960s in which differential methods (mainly for organic chemical species) received special attention from analysts. Indicative of this trend are a chapter on methods based on reaction rates by Lee in a series on *Organic Analysis* (20) and the textbook by Fritz and Hammond on quantitative organic analysis, in which reaction rate methods are discussed with emphasis on differential rate procedures (21).

The chapter by Lee, interestingly enough, mentions methods based on enzyme catalysis (including inhibition) and forecasts:

There is no doubt that methods based on catalysis by enzymes will increase in number and practicability as advances are made in the study of these substances and as more enzymes of known activity and purity become commercially available (21).

Time has proved this to be true. Although Stetter described applications of impure enzyme preparations to a variety of analytical problems (22), enzymes as analytical reagents (and clinical analysis as another recognized branch of chemical analysis) made their first impact in the mid-1960s and early 1970s. The 1970 fundamental review on kinetics in *Analytical Chemistry* also covered enzymatic methods. By this time the picture of chemical analysis as a branch of chemistry dominated by equilibrium-based determinations of inorganic species in mineral samples was fading away. Full utilization of a broad spectrum of chemical reactions characterizes this period. Typical examples are the recognition of chemical inhibition and activation of metal-catalyzed reactions as a means for determining a variety of organic chemical species (23, 24) and exploitation of catalysis for indicating titrimetric end points (25, 26), areas which received increased consideration in the 1960s.

Applications of catalytic determinations and less frequently of differential rate methods have extended through the mid-1960s, the 1970s, and are well established in the 1980s. The 1960s, however, brought another change: the availability of improved electronic devices which, with their inherent dependability, erased the argument that time was an elusive and unreliable variable to measure.

Several arguments have often been used to reject kinetic-based determinations. These arguments revolved around the following: (a) the imprecision of time measurements; (b) the apparent sophistication of kinetic discussions that led to the assumption that kinetic determinations are necessarily "too mathematically involved"; (c) the need for temperature control; (d) the need for means to quench reactions in order to perform measurements under steady-state conditions; and (e) the belief that kinetic procedures are time consuming because they would require careful preparation of solutions, close control of reaction conditions, and sufficient time for reaction so as to measure significant changes and offset the imprecisions introduced by imperfect control of parameters affecting the velocity of the reaction. All these problems started to be solved with better instrumentation and data handling; to day they are under control, and so the objections can be rather easily discarded. The progress introduced in the 1960s can be traced back to the "operational amplifier revolution," which is well depicted by the now classic book on practical electronics by Malmstadt, Enke, and Toren (27), but for which the road was paved 20 years earlier by another classic written by Müller, Garman, and Droz (28).

In the 1950s a great deal of activity was dedicated to developing recording

**Table 1.1. Selected Instrumental Developments in Automated Reaction-Rate Methodology (in chronological order)**

| Instrumental Development | Author(s) and Reference |
| --- | --- |
| Major objective to assemble an instrument from commercially available basic units and interface it with a control system and enzyme injector; control system in turn connected to the control unit. The commercially available unit was the Spectro section of the Spectro-Electro titrator (E. H. Sargent & Co.). The analytical aim was to shorten to roughly 1 min the clinical determination of glucose based on glucose oxidase catalysis and monitoring of released $H_2O_2$ by oxidation of a dye. | H. V. Malmstadt, and G. P. Hicks, *Anal. Chem.*, **32**, 394–398 (1960). |
| Automatic, potentiometric, reaction rate setup to measure changes in the emf of a concentration cell. Inspired by the earlier development of Malmstadt and Hicks (1960), it introduced extensive modifications and demonstrated the applicability of the variable-time procedure to nonlinear response curves. | H. V. Malmstadt, and H. L. Pardue, *Anal. Chem.*, **33**, 1040–1047 (1961); *Clin. Chem.*, **8**, 606–615 (1962). |
| First implementation of unsegmented continuous-flow sample processing for the measurement of the rate of an enzyme-catalyzed reaction in a continuous fashion. | W. J. Blaedel, and G. P. Hicks, *Anal. Chem.*, **34**, 388–394 (1962). |
| Automatic determination of the slope of the rate curve near zero reaction time. The measurement was accomplished by matching the slope of the output signal from an intergrator circuit with the slope of the signal from the chemical system. | H. L. Pardue, *Anal. Chem.*, **36**, 633–636 (1964). |
| Adaptation of an electrical divider for the continuous computation of reciprocal time in combination with automatic control equipment to provide digital readout of concentration. | H. L. Pardue, C. S. Frings, and C. J. Delaney, *Anal. Chem.*, **37**, 1426–1429 (1965). |
| Reliable and inexpensive pH-stat modular unit with digital readout built around the Heath EUW-301 pH recording electrometer. | H. V. Malmstadt, and E. H. Peipmeier, *Anal. Chem.*, **37**, 34–44 (1965). |
| Automatic digital printout of concentration by an instrument that measures time between two preset points on the response curve (variable-time procedure). | H. L. Pardue, and R. L. Habig, *Anal. Chim. Acta*, **35**, 383–390 (1966). |

8

H. V. Malmstadt, and S. R. Crouch, *J. Chem. Ed.*, **43**, 340–353 (1966).

Presentation of three modular rate measurement systems based on comparison (slope matching). The evolution of feedback control is nicely illustrated first with manual balancing, second with an electromechanical servo loop, and finally with an all-electronic feedback loop.

R. H. Stehl, D. W. Margerum, and J. J. Latterell, *Anal. Chem.*, **39**, 1346–1350 (1967).

Instrument combining rapid mixing techniques with automated measurement techniques previously developed by Malmstadt et al. (1966) and Pardue et al. (1965).

G. E. James, and H. L. Pardue, *Anal. Chem.*, **40**, 796–802 (1968).

All-electronic reciprocal-time analog computer. All-solid-state circuitry. Response considerably faster than for electromechanical systems used previously.

G. E. James, and H. L. Pardue, *Anal. Chem.*, **41**, 1618–1623 (1969).

Interfacing of a small general-purpose digital computer for on-line processing of reaction rate data. The instrumental setup includes a photometric system employing optical feedback to yield photometric drift of less than 0.02% transmittance per hour.

A. C. Javier, S. R. Crouch, and H. V. Malmstadt, *Anal. Chem.*, **40**, 1922–1925 (1968); **41**, 239–243 (1969).

First analytical application (other than biochemical) of stopped-flow mixing.

S. R. Crouch, *Anal. Chem.*, **41**, 880–883 (1969).

Computation of the reciprocal of time by measuring the period of a pulse train whose frequency is directly proportional to the desired time interval. Hybrid, analog–digital system.

R. A. Parker, H. L. Pardue, and B. G. Willis, *Anal. Chem.*, **42**, 56–61 (1970).

Utilization of all-digital circuitry to generate a frequency proportional to reciprocal time and concentration.

J. D. Ingle, Jr., and S. R. Crouch, *Anal. Chem.*, **42**, 1055–1060 (1970).

Fixed-time digital timing system for computing initial rates by a digital integration procedure.

**Table 1.1.** (*continued*)

| Instrumental Development | Author(s) and Reference |
|---|---|
| Detailed discussions on instrumental design of a spectrophotometric system for analytical applications of kinetic methods. | T. E. Weichselbaum, W. H. Plumpe, Jr., and H. B. Mark, Jr., *Anal. Chem.*, **41**, 103A–107A (1969). |
| | T. E. Weichselbaum, W. H. Plumpe, Jr., R. E. Adams, J. C. Hagetty, and H. B. Mark, Jr., *Anal. Chem.* **41**, 725–736 (1969). |
| Computer-controlled instrument system by which routine as well as decision-making procedures of data interpretation and experimental design are automated. | S. N. Deming, and H. L. Pardue, *Anal. Chem.*, **43**, 192–200 (1971). |
| Computer system designed to perform kinetic experiments to desired end points, making all necessary decisions and controlling all instrumentation by means of a special-purpose time-sharing system. | A. A. Eggert, G. P. Hicks, and J. E. Davis, *Anal. Chem.*, **43**, 736–747 (1971). |
| Simple switching networks for variable-time, catalytic end-point indication, and for collecting information during the inhibition portion of catalytic titrations as well as the catalytic part. | H. A. Mottola, *MPI Appl. Notes* **6**, 17–21 (1971). |
| Completely automatic stopped-flow spectrophotometer. | P. M. Beckwith, and S. R. Crouch, *Anal. Chem.*, **44**, 221–227 (1972). |
| Completely automated, computer-controlled spectrophotometric system incorporating a reagent preparation system more flexible and less subject to mechanical problems than syringe-based systems. | G. E. Mieling, R. W. Taylor, L. G. Hargis, J. English, and H. L. Pardue, *Anal. Chem.*, **48**, 1686–1693 (1976). |
| Instrumentation for kinetic nephelometric determination. Application to rapid analysis for specific serum proteins by immunoprecipitin reactions. | J. C. Sternberg, *Clin. Chem.*, **23**, 1456–1464 (1977). |

10

Design and operation of a fluorometric reaction rate instrument.

R. L. Wilson, and J. D. Ingle, Jr., *Anal. Chem.*, **49**, 1060–1065 (1977).

Computer-controlled bipolar pulse conductivity system specifically developed as a detector for chemical rate determination.

K. J. Caserta, F. J. Holler, S. R. Crouch, and C. G. Enke, *Anal. Chem.*, **50**, 1534–1541 (1978).

Microcomputer-automated on-line reagent dilution system that considerably reduces the time required for reagent preparation prior to stopped-flow mixing.

S. Stieg, and T. A. Neiman, *Anal. Chem.*, **52**, 798 (1980).

Microcomputer-automated and -controlled intensified diode array system capable of acquiring a chemiluminescence spectrum from 200 to 840 nm in 4 ms under direct memory control.

D. F. Marino, and J. D. Ingle, Jr., *Anal. Chem.*, **53**, 845 (1981).

Computer-automated unsegmented continuous-flow system providing convenient reagent conservation. All elements use computer feedback control for stability and application.

D. J. Hooley, and R. E. Dessey, *Anal. Chem.*, **55**, 313 (1983).

11

**Table 1.2. Some chronological antecedents to the Historical Developments in Kinetic-Based Determinations Singled out for Their Effect on Evolution of the Area**

| Year | Event | Author(s) | Reference |
|------|-------|-----------|-----------|
| 1876 | Report of the catalytic effect of vanadium on the chlorate oxidation of aniline. | Guyard | 6 |
| 1885 | Estimation of vanadium based on its catalytic effect on the oxidation of aniline to aniline black. | Witz and Osmond | 7 |
| 1921 | Characterization of three tannin fractions by analysis of the rate of oxidation of these substances on a kinetic basis. | Clarens | 19 |
| 1934 | Proposal to determine iodide based on its catalysis of the As(III) oxidation by Ce(IV). | Sandell and Kolthoff | 8 |
| 1937 | Specific conditions given for the determination of iodide based on its catalysis of the Ce(IV)–As(III) reaction. | Sandell and Kolthoff | 9 |
| 1947 | Postulation of the single-point method for resolving binary mixtures in differential rate methods. | Kolthoff and Lee | 15 |
| 1959 | Resolution of mixtures using logarithmic extrapolation methods (as applied to the iodometric determination of ozone and organic oxidants). | Saltzman and Gilbert | 29 |

12

| Year | Event | Author(s) | Reference |
|---|---|---|---|
| 1960 | Beginning development of instrumental automation for kinetic determinations. | Malmstadt and Hicks | See Table 1.1 |
| 1962 | Method of proportional equations applied to the kinetic analysis of mixtures. | Garmon and Reilley | 30 |
| 1964 | First review totally dedicated to kinetics in analytical chemistry in the journal *Analytical Chemistry*. | Rechnitz | See Figure 1.1 |
| 1965 | Symposium on the application of kinetics to analytical problems during the 150th National Meeting of the American Chemical Society (Atlantic City, N.J., September 12–17, 1965). | | |
| 1966 | Publication of the first monograph on kinetic methods of determination, with emphasis on catalytic determinations. | Yatsimirskii | 12 |
| 1968 | Publication of the second monograph on kinetic aspects of analytical chemistry. | Mark and Rechnitz | 18 |
| 1983 | First International Symposium on Kinetics in Analytical Chemistry (Cordoba, Spain, September 27–30, 1983). | | |

spectrophotometers and attachments to convert nonrecording single-beam instruments into recording double-beam units. The capability for continuous recording of a signal that relates to concentration changes as a chemical system comes to equilibrium was undoubtedly the first significant step in the instrumental promotion of measurements under dynamic conditions in systems approaching equilibrium. Modular instrumentation received renewed impetus in the 1960s and together with the strip chart, potentiometric recorder signal-conditioning modules appeared in most academic analytical surroundings. The instrumental impact on reaction rate methodology was thus another revolution within the 1960s. Table 1.1 summarizes those developments considered of interest in the evolutionary process by which instrumentation and computer applications made kinetic measurements competitive with (and sometimes superior to) equilibrium-based measurements. These instrumental developments freed the chemist from the need for stopping the reaction or withdrawing samples for determination, and they erased in large portion objections a, c, d, and e listed earlier. Objection b, the apparent sophistication and mathematical involvement of kinetic-based determinations, still remains as a roadblock in some practicing quarters. Probably this explains what otherwise seems a puzzling neglect by instrument makers in recognizing kinetic-based methodology even when the unit being promoted is performing kinetic functions. Table 1.1 shows the evolution from the use of servo units of an electromechanical nature to all-electronic systems, from analog to hybrid, and from semiautomation to fully automated, controlled systems. Table 1.2 closes this historical overview of kinetic measurements in analytical chemistry by listing some chronological developments demarking salient steps in the evolution of this area.

## 1.2.  CHEMICAL REACTIONS AND PHYSICAL PROCESSES IN ANALYTICAL CHEMISTRY

Chemical reactions in solution represent the origin and justify the name of the subdiscipline we know as analytical chemistry. Their study and applications constitute a large portion of today's tasks of analytical nature. Environmental problems (requiring, e.g., water analysis), clinical chemistry, and many industrial analytical methods exploit the information generated by measuring a property of chemical reactions. Equally important, however, is the fact that many other analytical methods and techniques are based purely on the measurement of a physical property of the system or chemical species of interest. In either case—whether chemical measurement or physical measurement—a rich store of information is disclosed by observation and kinetic measurement in chemical reactions or physical processes taking place in analytical procedures.

## .1.2.1.  Chemical Reactions

As already stated, the bulk of analytical chemistry based on chemical reactions centers on solution chemistry, particularly aqueous solutions. The basic types of reactions used for determinative purposes encompass the traditional four in equilibrium-based measurements: precipitation (ion exchange), acid–base (proton exchange), redox (electron exchange), and complexation (ligand exchange). These four basic types, or cases that can be reduced to them, are also found in kinetic-based measurements with some distinguishable trends. Redox reactions are employed mainly in catalytic methods of nonenzymatic nature; ligand exchange reactions have been predominant in differential rate methods. Proton transfer and precipitation reactions, on the other hand, have had a much more limited use in kinetic-based determinations. Table 1.3 itemizes the chemistry involved in commonly encountered kinetic-based measurements for analytical purposes (31).

Table 1.3. Classification of Kinetic Methods Based on Chemistry of Reactions Employed

---

*Homogeneous Systems*

Catalytic Methods
  Enzymatic methods employing soluble enzyme preparation.
  Nonenzymatic methods employing mainly catalysis of redox reactions by
    transition metal ions.
Uncatalyzed reaction rate methods
  Single-component determinations.
  Multicomponent determinations (differential reaction rate methods).

*Heterogeneous Systems*

Kinetic methods based on electrode reactions
Enzymatic methods employing immobilized enzyme preparations

---

The rate profile indicated in Figure 1.3 is exhibited by all types of chemical reactions regardless of their complexity. The kinetic and equilibrium regions illustrated in this figure reflect the two complementary approaches that contemporary analytical chemists apply: (a) signal measurements made in systems at equilibrium (thermodynamic approach), and (b) signal measure-

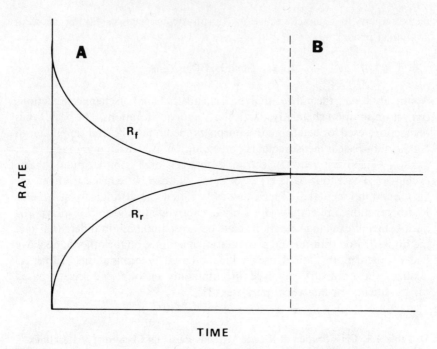

**Figure 1.3.** Reaction rate–time profile. $R_f$, forward rate, $R_r$, reverse rate, (A) Kinetic region, (B) Equilibrium region ($R_f = R_r$).

ments made under dynamic conditions in systems approaching equilibrium (kinetic approach). Measurements in the kinetic region are affected by a variable normally of no consequence in the equilibrium region: time. Correlation of time and the physicochemical property or properties of the system under observation can be reliably controlled with modern instrumentation. However, the time at which the equilibrium region is reached is critical for the mixing approach used in bringing the reactants into reaction. In a loose manner chemists talk of fast (or rapid) and slow reactions. But how fast must a reaction be in order to qualify for the "fast" or "slow" group? Could we use some rate parameter to classify reactions in such a way? Values of rate coefficients cannot provide a reliable criterion because they change with temperature and experimental conditions, and rates themselves can be changed by changing concentrations of the reactant(s). The half-life of a reaction is commonly used as a defining parameter to classify reactions as fast or slow in a rough, arbitrary way. For instance, the literature contains statements such as this:

We shall find it useful to adopt the pragmatic definition that a fast reaction is one which has a half-life that is small compared with the response time of humans, for example on the order of seconds, when the reactants are present in concentrations usually employed for chemical reactions (e.g. 0.10 $M$) and the temperature is room temperature (32).

Adopting this definition, which includes reactions with complicated mechanisms and isolated elementary reactions and which is keyed to the experimental techniques used to follow the reaction, one can state that reactions with half-times larger than about 10 s are considered slow, whereas those with half-times smaller than 10 s are considered fast (33). The 10 s time is, however, qualified by the *about*, and there is nothing binding about it but the vague boundary dictated by the type of analytical approach used to follow the reaction. Also worth mentioning that the half-time is independent of the concentrations of both reactants only for first-order reactions, moreover, it is always dependent on the rate coefficient for the reaction under consideration, as can be seen for even the simple case of a reaction involving a single reactant ($n$A→products) for which the rate is in general

$$\frac{d[A]}{dt} = k([A]_0 - [A])^n$$

which after integration gives

$$t_{1/2} = \frac{v}{k} \frac{1}{[A]_0^{n-1}}$$

where $n$ = order of the reaction, $[A]_0$ = initial concentration of the reactant, $k$ = corresponding rate coefficient, $t_{1/2}$ = half-time, and $v$ = numerical quantity depending on the order of the reaction. Thus the same objection to using rate coefficients or just simple rates could be used when the value of the half-time is used to classify reactions as slow or fast. The arbitrary acceptance of a 10-s boundary can be equally questioned; although this boundary is quite satisfactory for most analytical methods of determination, observations can be made today in favorable cases, even on a time scale of picoseconds (34).

Qualifications aside, what can be learned from these considerations is the relative meaning of the term *fast* as applied to chemical reactions and the recognition that the dictating factor is not the speed of the reaction but the time window provided by the mixing of reactants and the observation of reaction progress. Realizing this, one can easily understand why even typical monographs on "fast" reactions (33, 34) acknowledge that treatment of the topic itself is dictated by experimental innovations and why they organize the subject according to type of experimental technique used rather than kinetic behavior of the systems chosen for illustration. Table 1.4 lists the three basic approaches taken with respect to observing chemical reactions and physical processes to obtain kinetic information. Slow reactions are generally observed by either a batch or a continuous-flow operation after reactant mixing. Fast reactions, on the other hand, are followed by a larger variety of experimental approaches. From a determinative analytical viewpoint, stopped-flow mixing, continuous-flow methods, and some spectral phenomena (e.g., fluorescence and phosphorescence) have received the most attention among approaches utilizing fast reacting systems.

**Table 1.4. Classification of Experimental Approaches**

*Methods based on reactant mixing and following of reaction course as system approaches equilibrium*

1. Batch mixing
2. Stopped-flow mixing
3. Continuous-flow mixing

*Relaxation methods*

(system originally at equilibrium is suddenly disturbed
    from the equilibrium position and observed as it relaxes
    back to such a position)
1. Jump techniques
    a. Temperature-jump
    b. Pressure-jump
    c. Electric impulse
    d. Photon impulse
2. Periodic relaxation techniques
    a. Ultrasound wave propagation
    b. Dielectric relaxation
3. Spectral excitation techniques
    a. Fluorescence
    b. Phosphorescence

*Equilibrium methods*

(including a few techniques that permit extracting rate
    data of some process without disturbing the system at
    equilibrium)
1. Resonance techniques
    a. Nuclear magnetic resonance
    b. Electron paramagnetic resonance

The ultimate goal of analytical chemistry is to be able to characterize and measure the composition of matter in its different physical states. Determinations per se occupy just one of the provinces of analytical chemistry. Sampling, sample processing, preliminary chemical reactions, and separations are areas of analytical chemistry in which the rates of chemical processes may need to be known. Although the relevance of kinetics may seem more incidental in these cases than in kinetic-based determinations, it is not at all incidental. The rates of reactions leading to contamination or to loss of analyte are relevant in sampling, in sample preservation, and in sample transport. The rates of side reactions leading to the formation of interferent species, or to masking of the species of interest, are relevant in choosing, the processing approach. For example, implementing a continuous-flow processing system with kinetic measurement under stopped-flow conditions practically eliminates interferences encountered in the determination of Cu(II) in biological samples by its catalytic effect on the 2,2'-dipyridyl ketone hydrazone–$H_2O_2$ reaction (35). Another illustration is the finding that, at the pressures encountered in flameless atomizers for atomic absorption measurements, reaction kinetics plays a significant role in predicting the degree to which the analyte and an interferent substance may interact (36).

### 1.2.2. Physical Processes

As illustrated above, the rate of change of chemical composition with time is clearly the basis for kinetic-based determinations. Equally important but not as easily identified are rates of change of physical properties of a system (37). Although it receives some attention when considering the fundamentals of some analytically relevant processes (e.g., electrode processes and chromatography), the value of diffusion as rate information is generally diluted or hidden. Diffusion as a time-dependent process of equalizing concentration and convection as another time-dependent process of equalizing density and temperature play key roles in the performance of continuous-flow sample–reagent(s) processing systems (38). Many analytical techniques or approaches, if scrutinized closely, are perceived to be of kinetic nature or to involve key kinetic components. Table 1.5 singles out some of these techniques to illustrate the impact, often taken for granted, that physically based time-dependent processes have on analytical chemistry as we know it today. Pertinent references are included.

**Table 1.5. Selected Areas of Analytical Interest in Which Kinetics Play Direct or Indirect Important Roles**

---

### Formation of Precipitates from Solution

Of obvious direct interest in gravimetric determination. Theoretical aspects of precipitation are derived from accepted theories of nucleation and crystal growth. Details on the kinetics of nucleation as well as diffusion-controlled growth, nucleation-controlled growth, and dislocation-controlled growth are discussed in detail by Nielsen (39).

### Crystallization

As indicated above the kinetics of crystal growth is of interest in gravimetry. Theories of nucleation and crystal growth also relate to purification techniques such as zone melting (40) and, for instance, whiskers growth for chromatography (41). Most information catalogued for whisker growth on metallic surfaces seems to apply to silica whisker growth (42).

### Solvent Extraction

Although majority of solvent extraction processes of analytical interest are sufficiently fast, some are slow. In most cases kinetics in these processes involves rate of formation of the extractable species as well as rate of mass transfer across the interface. Solvent extraction has been shown to be a rather simple and efficient manner to obtain rate information on metal chelate formation even with fast reacting systems (43). Detailed review on kinetics of solvent extraction of metal ions is available (44).

### Ion Exchange

Experimental conditions affect rate of ion exchange (45) by altering either particle diffusion, film diffusion, or both, and are equally well applicable to immobilized chelating agents (46).

### Elution Chromatography

Chromatographic separations are nonequilibrium-driven processes. The introduction to a generalized theory by Giddings (47) provides a clear understanding of the factors (kinetic in nature) contributing to band spreading.

### Electrode Kinetics

Current–potential dependence of electrode processes provides characteristic information on electrode kinetics, which depends on rate of electron transfer at the interface as well as pure chemical kinetics in and in the vicinity of the double layer. Subject has been reviewed in depth (48, 49, 50).

**Table 1.5.** (*continued*)

*Thermal Methods*

Direct injection enthalpimetry is perhaps the thermal method in which the kinetic nature is most obvious since it is based on monitoring change in temperature with time (51). Kinetics is also entrenched in other thermal methods, as can be seen by the theoretical foundations of differential thermal analysis such as the Newton cooling equation and the rate of heat production in chemical processes (52).

*Continuous-Flow Systems*

Kinetics is at the heart of continuous-flow analyzers. Credit for the recognition of the kinetic nature of signal characteristics in air-segmented continuous-flow systems must be given to Thiers et al. (53). Details on kinetic factors that determine peak characteristics in such systems can be found in specialized monographs (54). Unsegmented continuous-flow systems, broadly known as flow injection systems, are even richer in kinetic content (38, 55, 56). In these procedures the shape and quality of the analytical signal is directly influenced by dispersion. Dispersion occurs as a result of mass transfer and mixing during sample–reagent(s) transport by the carrier stream to a detector. The occurrence of chemical reactions during the transport can alter diffusion and convection processes (55, 56). A chemical kinetic model based on series reactions has been proposed and tested for the description of the transient signal obtained when a sample is injected directly into a well-stirred detection zone as part of a closed-loop unsegmented continuous-flow system (57).

*Atomic Absorption Spectrometry*

A series process model in atomic absorption spectrometry with a graphite furnace was proposed by Fuller (58). A generalized theory considering both diffusion and reaction kinetics in analytical flame spectroscopy was advanced by Li (59). The model was further developed to study chemical interferences in atomization at various dissociation and vaporization rates (60).

# REFERENCES

1. F. Szabadvary, *History of Analytical Chemistry*, Pergamon, Oxford, 1966.
2. E. B. Sandell and P. J. Elving, *Treatise on Analytical Chemistry*, Part I: *Theory and Practice*, Vol. 1, I. M. Kolthoff and J. P. Elving, Eds., Interscience, New York, 1959, Chapter 1.
3. I. M. Kolthoff, E. B. Sandell, E. J. Meehan, and S. Bruckenstein, *Quantitative Chemical Analysis*, 4th ed., Macmillan, Toronto, 1969, Chapter 10.

4. H. H. Willard and N. K. Tang, *J. Am. Chem. Soc.*, **59**, 1190 (1937).

5. L. Gordon, M. L. Salutsky, and H. H. Willard, *Precipitation from Homogeneous Solution*, Wiley, New York, 1959.

6. M. A. Guyard, *Bull. Soc. Chim. Paris*, **25**, 58 (1876).

7. C. Witz and F. Osmond, *Bull. Soc. Chim. Paris*, **45**, 309 (1885).

8. E. B. Sandell and I. M. Kolthoff, *J. Am. Chem. Soc.*, **56**, 1426 (1934).

9. E. B. Sandell and I. M. Kolthoff, *Mikrochim. Acta*, 9 (1937).

10. L. Szebelledy and M. Ajtai, *Mikrochemie*, **25**, 258 (1938).

11. L. Szebelledy and M. Ajtai, *Mikrochemie*, **26**, 87 (1939).

12. K. B. Yatsimirskii, *Kinetic Methods of Analysis*, Pergamon, Oxford, 1966.

13. H. A. Mottola and H. B. Mark, Jr., *Anal. Chem.*, **54**, 62R (1982).

14. P. W. West, *Anal. Chem.*, **26**, 121 (1954).

15. I. M. Kolthoff and T. S. Lee, *J. Polym. Sci.*, **2**, 200 (1947).

16. I. M. Kolthoff, T. S. Lee, and M. A. Mairs, *J. Poly. Sci.*, **2**, 220 (1947).

17. S. Siggia, *Quantitative Organic Analysis via Functional Groups*, Wiley, New York, 1949.

18. H. B. Mark, Jr., and G. A. Rechnitz, *Kinetics in Analytical Chemistry*, Wiley-Interscience, New York, 1968.

19. M. J. Clarens, *Bull. Soc. Chim. Fr.*, **29**, 837 (1921).

20. T. S. Lee, in *Organic Analysis*, Vol. 2, J. Mitchell, Jr., I. M. Kolthoff, E. S. Proskauer, A. Weissberger, Eds., Interscience, New York, 1954, p. 237.

21. J. S. Fritz and G. S. Hammond, *Quantitative Organic Analysis*, Wiley, New York, 1957, Chapter 9.

22. H. Stetter, *Enzymatische Analyse*, Verlag Chemie, Weinheim, N.Y. (1951).

23. H. A. Mottola, *Anal. Chim. Acta*, **71**, 443 (1974).

24. D. P. Nikolelis and T. P. Hadjiioannou, *Rev. Anal. Chem.*, **4**, 81 (1979).

25. H. A. Mottola, *Talanta*, **16**, 1267 (1969).

26. T. P. Hadjiioannou, *Rev. Anal. Chem.*, **3**, 82 (1976).

27. H. V. Malmstadt, C. G. Enke, and E. C. Toren, Jr., *Electronics for Scientists*, Benjamin, New York, 1962.

28. R. H. Müller, R. L. Garman, and M. E. Droz, *Experimental Electronics*, Prentice-Hall, New York, 1942.

29. B. E. Saltzman and N. Gilbert, *Anal. Chem.*, **31**, 1914 (1959).

30. R. G. Garmon and C. N. Reilley, *Anal. Chem.*, **34**, 600 (1962).

31. H. A. Mottola and H. B. Mark, Jr., "Kinetic Methods," in *Instrumental Analysis*, G. D. Christian, and J. E. O'Reilly, Eds., 2nd ed., Allyn and Bacon, Boston, 1985, Chapter 18, pp. 560–593.

32. W. C. Gardiner, Jr., *Rates and Mechanisms of Chemical Reactions*, Benjamin, New York, 1969, p. 226.

33. E. F. Caldin, *Fast Reactions in Solution*, Wiley, New York, 1964.

34. J. N. Bradley, *Fast Reactions*, Clarendon Press, Oxford, 1975.

35. F. Lazaro, M. D. Luque de Castro, and M. Valcarcel, *Anal. Chim. Acta*, **165**, 177 (1984).

36. J. A. Holcombe, R. H. Eklund, J. E. Smith, *Anal. Chem.*, **51**, 1205 (1979).

37. H. A. Mottola, *J. Chem. Educ.*, **58**, 399 (1981).

38. C. C. Painton and H. A. Mottola, *Anal. Chim. Acta*, **154**, 1 (1983).

39. A. E. Nielsen, *The Kinetics of Precipitation*, Macmillan, New York, 1964.

40. H. Schildknecht, *Zone Melting*, Verlag Chemie–Academic Press, Weinheim, N.Y., 1966.

41. J. D. Schiecke, N. R. Comins, and V. Pretorius, *J. Chromatogr.*, **112**, 97 (1975).

42. R. F. Strickland-Constable, *Kinetic Mechanisms of Crystallization*, Academic Press, New York, 1968.

43. H. Freiser, *Acc. Chem. Res.*, **17**, 126 (1984).

44. P. R. Danesi and R. Schiarizia, *CRC Crit. Rev. Anal. Chem.*, **10**, 1 (1980).

45. F. Helfferich, *Ion Exchange*, McGraw-Hill, London, 1962.

46. M. A. Marshall and H. A. Mottola, *Anal. Chem.*, **57**, 729 (1985).

47. J. C. Giddings, *Dynamics of Chromatography*, Part I: *Principles and Theory*, Dekker, New York, 1965.

48. P. Delahay, *Double Layer and Electrode Kinetics*, Wiley-Interscience, New York, 1965.

49. K. J. Vetter, *Electrochemical Kinetics (Theoretical and Experimental Aspects)*, Academic Press, New York, 1967.

50. H. R. Thirsk and J. À. Harrison, *A Guide to the Study of Electrode Kinetics*, Academic Press, London, 1972.

51. J. Barthel, *Thermometric Titrations*, Wiley-Interscience, New York, 1975, pp. 66–70.

52. M. I. Pope and M. D. Judd, *Differential Thermal Analysis*, Heyden, Philadelphia, 1977.

53. R. E. Thiers, R. R. Cole, and W. J. Kirsch, *Clin. Chem.* (Winston-Salem, N.C.), **13**, 451 (1967).

54. W. H. C. Walker, in *Continuous Flow Analysis: Theory and Practice*, W. B. Furman, Ed., Dekker, New York, 1976, Chapter 7.

55. C. C. Painton and H. A. Mottola, *Anal. Chim. Acta*, **158**, 67 (1984).

56. D. Betteridge, A. P. Wade, and C. Z. Marczewski, *Anal. Chim. Acta*, **165**, 227 (1984).

57. H. A. Mottola and A. Hanna, *Anal. Chim. Acta*, **100**, 167 (1978).

58. C. W. Fuller, *Analyst (London)*, **99**, 739 (1974).

59. K.-P. Li, *Anal. Chem.*, **53**, 317 (1981).

60. K.-P. Li and Y.-Y. Li, *Anal. Chem.*, **53**, 2217 (1981).

# CHAPTER

## 2

# CATALYTIC METHODS: NONENZYMATIC METHODS

Before discussing catalytic methods in detail, the term *catalyst* must be defined unambiguously. This is needed because the category "catalytic methods" denotes all chemical determinations based on monitoring the rate of a catalyzed reaction. The criteria for catalysis as well as the name itself (*catalysis* comes from the Greek meaning "loosen") were originated by J. J. Berzelius around 1835–1836 (1). Berzelius's definition indicates that (a) the catalyst is unchanged chemically at the end of the reaction; (b) a small amount of catalyst is often sufficient to bring about a considerable amount of reaction; and (c) the catalyst does not affect the position of equilibrium in a reversible reaction. These concepts lead to the definition of a catalyst as a substance that lowers the free energy of activation of any change that can occur with a diminution of free energy. Lowering the free energy of activation occurs through a reaction path by which the catalyst is constantly regenerated (the catalytic cycle) so that for all practical purposes the concentration of the catalyst remains constant. A catalyst can only increase the rate of reaction, so use of terms such as *negative catalyst* for substances that increase the free energy of activation should be abandoned.

It should be pointed out here that a catalyst must influence the forward and reverse reaction rates in the same proportion (2). Therefore, the last few words of our definition are not essential, because increasing the rate of a nonspontaneous process is of no interest to the analyst. The key conceptual role of the catalytic cycle is clearly embedded in this defining view of catalysis.

A variety of catalytic effects have been employed in catalytically based analytical determinations; the most important ones are listed in Table 2.1. Catalytic determinations can be broadly viewed as involving two main approaches:

1. Direct use of primary catalytic effects (determination of catalyst).
2. Use of modified catalytic rates (indirect determination of modifier).

Catalytic rate modifications included in the second area are inhibition and activation, the latter resulting in modified true catalytic paths or promoting effects.

**Table 2.1. Types of Catalytic Effects Used in Kinetic-Based Analytical Determinations**

*A. Homogeneous Catalysis*

1. *Oxidation–reduction catalysis.* By far the most common of catalytic effects utilized in analytical determinations. Most cases involve catalysis by transition metal ions, but catalysis by anions is also represented.
2. *Enzyme catalysis.* Much used in biomedical and clinical chemistry for the determination of enzymes, substrates, inhibitors, and activators. Application of enzymes to other areas of determination is receiving increasing interest.
3. *Catalyzed ion-exchange reactions.* Includes (a) monodentate ligand exchange catalyzed by metal ions, and (b) multidentate ligand exchange catalyzed by metal ions and/or ligands.
4. *Acid–base catalysis.* Offers poor selectivity.and limits of detection. Actually, many reactions catalyzed by hydronium ions are also catalyzed by some metal ions, etc. (Lewis acids). Application is limited to specialized cases[a] or to differential simultaneous determinations if the ratio of rate constants is very large.
5. *Exchange of oxidation states between ions of the same element.* Mechanistically resembles other redox reactions, but not numerous. Catalyzed by anions acting as "bridges" for electron exchange between two oxidation states of the same species. Radioactive labeling is used to monitor the catalytic effect (7, 8).

*B. Heterogeneous catalysis*

1. *Catalytic voltammetric currents.* Mechanistically resembles homogeneous redox catalysis, involving catalytic cycles before or after the electron-transfer process at the electrode surface. Although this process is heterogeneous, analytical applications center on determining bulk concentrations in solutions.
2. *Immobilized enzyme preparations.* Mostly applied as reactors for potentiometric detection ("enzyme electrodes") or in continuous-flow analyses.

---

[a] Some examples of particular cases of acid–base catalytic applications in analytical chemistry can be mentioned: (a) titrations of 2,6-disubstituted phenols, keto–enol mixtures, imides, and traces of acid, with nonaqueous alkaline solutions (3); (b) titration of tertiary amines and salts of organic acids in acetic acid (4, 5, 6).
Reprinted with permission from H. A. Mottola, *CRC Crit. Rev. Anal. Chem.,* **4,** 229 (1975). Copyright CRC Press, Inc., Boca Raton, FL.

## 2.1. MATHEMATICAL BASIS OF CATALYTIC METHODS BASED ON PRIMARY CATALYTIC EFFECTS

Two facts must be accounted for to develop mathematical relationships valid for catalytic determinations: (a) the uncatalyzed reaction proceeds simultaneously with the catalyzed reaction, and (b) the rate of the catalyzed reaction is directly or nearly proportional to the concentration of catalyst as a result of the catalytic cycle. In most instances of practical interest in· analytical chemistry, the first fact is of little significance because the catalyzed reaction is so much faster than the uncatalyzed one. The second fact is, however, the very basis of catalytic methods, and its mathematical modeling is the subject of this section.

Before entering into formulations, we must point out that the direct or nearly direct relationship between catalyst concentration and rate of catalyzed reaction is obeyed better if certain conditions are met, such as keeping at constant level all variables normally affecting the rate of chemical reactions (e.g., temperature, ionic strength, solvent used). Another important practical requirement needed for successful applications of conventional determinations based on primary catalytic effects is that concentrations of reactants, other than the catalyst and the species whose change in concentration is monitored, must be such as to make their effect on the rate pseudo-zero-order. The species whose change in concentration is being monitored is adjusted to first-order dependence. When all these facts and requirements are considered and met, for the generalized case

$$R + B \xrightarrow{C} P + Y \tag{2.1}$$

where R and B are reactants (R = monitored species), P and Y are products, and C is the catalyst, the following general expression can be written:

$$-\frac{d[R]}{dt} = k_u[R] + k_c[R][C]_0 = [R](k_u + k_c[C]_0) \tag{2.2}$$

in which $k_u$ = the rate coefficient for the uncatalyzed reaction (plus some concentration terms), $k_c$ = the rate coefficient for the catalyzed reaction (plus some concentration terms), and $[C]_0$ = the initial concentration of catalyst in the system.

Directing our attention to the term characterizing the catalyzed path in eq. 2.2, and accounting for the presence of the catalytic cycle, we can use the simplified two-step reaction scheme shown below to develop mathematical relationships for boundary conditions that explain the direct or nearly direct relationship between catalytic rate and catalyst concentration:

$$R + C \underset{k_{-1}}{\overset{k_1}{\rightleftharpoons}} (RC) + Y \tag{2.3}$$

$$(RC) + B \xrightarrow{k_2} P + C \tag{2.4}$$

This simplified mechanism for the catalytic cycle leads to two boundary conditions: (a) a pre-equilibrium case, and (b) a steady-state situation, which boundary depends on the relative values of the corresponding rate coefficients for reactions represented by eqs. 2.3 and 2.4. The mathematical relationships sought hinge on finding values for the concentration of the intermediate species (RC) to substitute in rate expressions for the rate-determining step.

If $k_1 \ll k_1$ and $k_2 \ll k_1$, the reaction represented by eq. 2.4 becomes rate determining and we have the *pre-equilibrium condition*. Under these conditions the equilibrium concentration of the species (RC) can be obtained from the equilibrium concentration quotient of reaction 2.3:

$$[RC] = \frac{K[C][R]}{[Y]} \tag{2.5}$$

where $K = k_{-1}/k_1 = $ equilibrium concentration quotient for reaction 2.3. Let us define $[C]_0$ and $[R]_0$ as equilibrium concentrations at time $t = 0$, so that at any time $t$,

$$[R] = [R]_0 - [(RC)] \tag{2.6}$$

and

$$[C] = [C]_0 - [(RC)] \tag{2.7}$$

In catalytic methods, $[R]_0 \gg [C]_0$; (that is, the catalyst is determined at very low, substoichiometric levels of concentration). Therefore, $[R]$ can be considered equal to $[R]_0$ and hence

$$[(RC)] = \frac{K\{[C]_0 - [(RC)]\}[R]_0}{[Y]} \tag{2.8}$$

which, expanded and rearranged, becomes

$$[(RC)](K[R]_0 + [Y]) = K[C]_0[R]_0 \tag{2.9}$$

from which an expression for $[(RC)]$ can be obtained:

$$[(RC)] = \frac{K[C]_0[R]_0}{K[R]_0 + [Y]} \tag{2.10}$$

In the pre-equilibrium case, eq. 2.4 is rate limiting and consequently

$$\text{rate} = k_2[(RC)][B] \tag{2.11}$$

$$\text{rate} = k_2[B]\frac{K[C]_0[R]_0}{K[R]_0 + [Y]} \tag{2.12}$$

If measurements are made under conditions for which $[Y] \ll K[R]_0$ (easily met in catalytic systems because either Y is not formed at all or, since the catalyst concentration is relatively very low, its concentration is negligible), and $[B] = $ constant (e.g., for measurements at early stages of the reaction, as in initial rate conditions), then

$$\text{rate} = -\frac{d[R]}{dt} = k'_2[C]_0 \tag{2.13}$$

with $k' = k_2[B]$. The relationship arrived at in eq. 2.13 reflects the proportionality between the concentration of catalyst and the rate as used in direct analytical determinations utilizing primary catalytic effects.

If $k_2 \gg k_1 \gg k_{-1}$, reaction 2.3 becomes rate determining, $[(RC)]$ is small because of the substoichiometric amounts of catalyst, and a *steady-state condition* develops defined by the approximation

$$\frac{d[(RC)]}{dt} = 0 \tag{2.14}$$

The rate expression then becomes

$$\text{Rate} = k_1[R][C] - k_{-1}[(RC)][Y] \tag{2.15}$$

Since $[(RC)]$ cannot be measured experimentally, we must resort to the steady-state approximation to obtain an expression for it:

$$\frac{d[(RC)]}{dt} = 0 = k_1[R][C] - k_{-1}[(RC)][Y] - k_2[(RC)][B] \tag{2.16}$$

As before we substitute $[R]_0 - [(RC)]$ and $[C]_0 - [(RC)]$ for $[R]$ and $[C]$, respectively. Expanding, simplifying, and rearranging gives

$$[(RC)] = \frac{k_1[R]_0[C]_0}{k_1([R]_0 + [C]_0) + k_{-1}[Y] + k_2[B]} \tag{2.17}$$

We now rewrite eq. 2.15 as

$$\text{rate} = k_1[R]_0[C]_0 - k_1[(RC)]([R]_0 + [C]_0) - k_{-1}[(RC)][Y] \tag{2.18}$$

and replace $[(RC)]$ in eq. 2.18 by its expression in eq. 2.17; expanding and simplifying then results in the following:

$$\text{rate} = \frac{k_1 k_2 [B][R]_0 [C]_0}{k_1([R]_0 + [C]_0) + k_{-1}[Y] + k_2[B]} \tag{2.19}$$

Since $[C]_0 \ll [R]_0$ and $k_1[R]_0 \gg k_{-1}[Y]$, if as before $[B] = \text{constant}$, we have

$$\text{rate} = k_2'[C]_0 \tag{2.20}$$

where

$$k_2' = \frac{k_1 k_2 [B][R]_0}{k_1[R]_0 + k_2[B]}$$

The relationship arrived at in eq. 2.20 again demonstrates for analytical determinations the direct relationship between catalyst and rate.

If properly treated, both approximations—for equilibrium and for steady state—have different ranges of applicability with a common region. The interrelation between the two approximations has been treated in detail by Pyun (9).

## 2.2. PRACTICAL TREATMENT OF DATA IN CATALYTIC DETERMINATIONS

Two main approaches distinctively categorize the treatment of data in catalytic determinations. These approaches can be distinguished as *differential* (derivative) and *integral* methods and subdivided as follows:

Differential methods:  Initial rate measurements
                       Slope measurements
Integral methods:      Fixed-time measurements
                       Variable-time measurements
                       Method of tangents

Differential methods of *initial rate* nature may involve direct evaluation of $d[\text{signal}]/dt$ at a time for all practical purposes equal to zero; or at a time close to zero. They may apply either a fixed-time approach (measuring $\Delta[\text{signal}]$ at a finite but short $\Delta t$ close to zero), or a variable-time approach to measurement (measuring the $\Delta t$ needed to attain a finite but small $\Delta[R]$ close to $[R]_0$). In defining the tolerable amount of reaction for a rate measurement to qualify as initial, one faces the same ambiguity as in trying to define fast and slow reactions. It is recommended that no more than 2% reactant conversion be allowed to occur, although measurements are generally made in a window that extends to as much as 10% product formation. Initial rates (estimated as $d$(signal)/$dt$ or $\Delta$(signal)/$\Delta t$) are linearly related to the initial catalyst concentration $[C]_0$, as mathematically shown earlier. Advantages generally attributed to initial rate measurements include the following: (a) there is no appreciable contribution from the back reaction and minimal complications from side reactions, and (b) the rate can be considered zero-order with respect to reactants since their concentrations do not change appreciably. (This should result in an improvement in reproducibility.) In any event, it should be realized that the success of initial rate measurements depends greatly on the availability of sufficiently sensitive and reproducible monitoring of small concentration changes in reactants or products at the beginning of the reaction.

*Slope measurements*, involving evaluation of $d$(signal)/$dt$ or $\Delta$(signal)/$\Delta t$ at any given time during the reaction, are seldom used except in association with multipoint rate evaluations of the reaction profile.

Integral methods, as their name implies, are based on the integration of the corresponding rate expressions over a finite, constant, but not necessarily small time interval $t_2 - t_1$ ($t_1$ may be equal to zero). The bases of the *fixed-time* approach can be illustrated by considering in a generalized manner the integration of the rate expression given by eq. 2.2. Rearranging eq. 2.2, we can have

$$-\frac{d[R]}{[R]} = k_u \, dt + k_c [C]_0 \, dt \tag{2.21}$$

Integrating between $t_1$ and $t_2$ and $[R]_1$ and $[R]_2$ leads to

$$\ln \frac{[R]_1}{[R]_2} = \Delta t (k_u + k_c [C]_0) \tag{2.22}$$

where $\Delta t = t_2 - t_1$. If $\ln([R]_1/[R]_2) = \Delta^*[R]$, we may write

$$[C]_0 = \frac{\Delta^*[R]}{\Delta t \, k_c} - \frac{k_u}{k_c} \tag{2.23}$$

If $\Delta t =$ constant (fixed time), $\Delta^*[R]$ is directly proportional to the initial catalyst concentration $[C]_0$, thus providing the basis for constructing calibration curves. Within limited concentration ranges, direct plotting of $\Delta[R] = [R]_2 - [R]_1$ versus $[C]_0$ provides linear working curves.

If $\Delta^*[R]$ is constant from run to run and equal to $K^*$, we may write

$$\frac{1}{\Delta t} = \frac{k_c}{K^*}[C]_0 + \frac{k_u}{K^*} \qquad (2.24)$$

which indicates that plots of $1/\Delta t$ versus $[C]_0$ should be linear. Equation 2.24 provides the basis for the *variable-time* approach.

The *method of tangents* is an integral method based on kinetic plots. The dependence of reaction order on the monitored species dictates the kind of plot to use. Most catalytic determinations are run under conditions for which such a dependence is first-order. For first-order dependence, plots of log[R] versus time yield a family of straight lines whose slopes (generally designated as pseudo-first-order constants) are linearly related to $[C]_0$. One can plot those slopes, such that the ordinate is expressed as tangents of the angles formed by the $x$ or $y$ axis and the straight line characterizing the rate and the abscissa shows the catalyst concentration. Such plots provide calibration for the method of tangents, which is widely reported in the Russian literature.

Mathematical justification for the method of tangents can be provided in a generalized form by rearranging eq. 2.22 and setting $t_1 = 0$. Then

$$-\ln[R] = -\ln[R]_0 + k'(t) \qquad (2.25)$$

where $k' = k_u + k_c[C]_0$, and hence the linear dependence of $k'$ and $[C]_0$ as a basis for working curves.

Typical plots of data using the fixed-time method, the variable-time method, and method of the tangents are shown in Figures 2.1, 2.2, and 2.3 respectively. On the basis of a study of the theoretical and experimental factors influencing the accuracy of reaction rate measurements, Ingle and Crouch (12) concluded that the variable-time approach is superior in most cases in which nonlinear response curves are obtained. This approach is also preferable in the determination of catalysts (including enzymes). The variable-time approach should also give a wider range of concentrations amenable to determination since adherence to pseudo-zero-order kinetics is unnecessary.

The fixed-time method is recommended for first-order or pseudo-first-order reactions and determination of substrates in enzyme-catalyzed reactions (12). By applying the theory of propagation of errors, Holler et al. (13), demonstrated that instantaneous rate measurements (the derivative approach) at a time equal to the reciprocal of the first-order or pseudo-first-order rate coefficient yield reaction rate values that are essentially independent of changes in parameters that affect the value of the rate coefficient. This observation challenges the preferability of initial rate measurements.

The method of tangents fails if the kinetics of the reaction(s) involved is

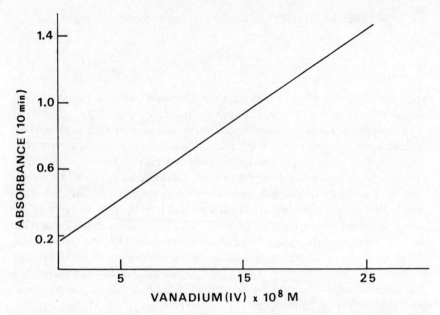

**Figure 2.1.** Typical fixed-time plot for the determination of vanadium(IV) using its catalytic effect on the *o*-dianisidine–-*t*-butyl hydroperoxide reaction in nonaqueous media. [Figure drawn using the equation $A_{10\,min} = (5.12 \times 10^6)[V(IV)] + 0.163$, given in the original paper, Ref. 10.]

complex and integration of the rate expressions is difficult or impossible. Since more measurements are performed by the method of tangents, its application is expected to result in better precision than use of the variable- or fixed-time methods (14).

There are a few more, much less frequently used approaches to measurement in catalytic determinations. Measurements based on the *length of induction periods* can be applied to some catalytic reactions in which there is a time delay between reactants mixing and appearance of change in the chemical composition of the system. Slow redox reactions involving halogens in various oxidation states, for instance, show induction periods. Landolt (15) was the first to observe this behavior, and the literature calls this type of measurements the Landolt effect or Landolt reactions. Svehla (16) has provided a good presentation of theoretical background and discussions on limits of detection, sensitivity, selectivity, and precision in this type of measurements. A generalized and simplified scheme for Landolt reactions can be written as

$$A + B \xrightarrow{k_1} I \tag{2.26}$$

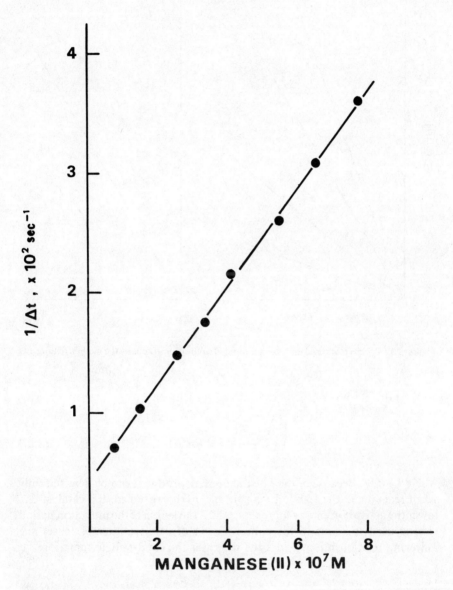

**Figure 2.2.** Variable-time plot for the determination of manganese(II) using the malachite green–periodate indicator reaction. [From Ref. 11.]

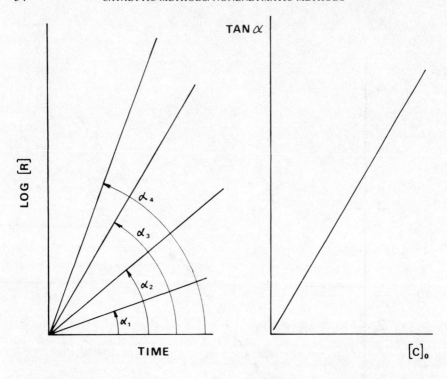

**Figure 2.3.** Typical plots used in the preparation of calibration graphs by the method of tangents.

$$I + L \xrightarrow{\quad k_2 \quad} P \text{ (or A)} \tag{2.27}$$

where $k_1 < k_2$. Because of the rate differential, product I can be detected only when reactant L (the Landolt reagent) has been consumed. If a catalyst can lower the activation energy for reaction 2.27, the length of the induction period can be used to estimate the concentration of catalyst. In most cases the following relationship provides the bases for the catalyst determination:

$$[C]_0 = \frac{1}{t_i} \tag{2.28}$$

where $t_i$ = length of induction period in time units.

An unconventional approach to measurement in catalytic determinations was introduced by Lopez-Cueto and Cueto-Rejon during the First International Symposium on Kinetics in Analytical Chemistry (17). They proposed to compare the signal–time profile for the sample with that of a reference solution to which the catalyst (analyte) species is added at a constant rate. Both signal–time profiles start at the same signal level at time zero. The reference curve will lag behind in rate with respect to the sample curve but will approach it, and at a given time $t_x$ the curves will cross. This time $t_x$ can be considered an "end-point" for the determination of catalyst. Beyond $t_x$, the rate of reaction in the reference solution exceeds that in the sample. Figure 2.4 illustrates these behaviors. At time $t_x$ the following expression applies:

$$[C]_x = [C]_s \left( 1 - \frac{V_0}{mt_x} \ln \frac{V_0 + mt_x}{V_0} \right) \tag{2.29}$$

where $[C]_x$ = the concentration of catalyst in the unknown (sample), $V_0$ = volume of reference solution, and $[C]_s$ = the concentration in the standard solution added at a constant rate $m$. Equation 2.29 permits direct calculation of the concentration of catalyst in the unknown sample without recourse to calibration curves. First-order or pseudo-first-order conditions need to prevail for eq. 2.29 to be valid. Alternatively, if dilution effects are

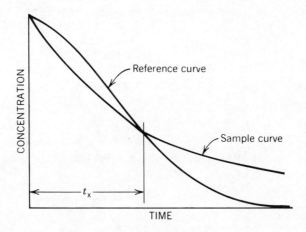

**Figure 2.4.** Sample and reference curves (theoretical) for catalytic determination by continuous standard addition to a reference solution. [G. Lopez-Cueto and A. Cueto-Rejon, Poster P.I.3, First International Symposium on Kinetics in Analytical Chemistry, Sept. 27–30, 1983. Reproduced with permission.]

negligible, a plot of $t_x$ versus $[C]_x$ may be used since the equation simplifies to

$$[C]_x = \frac{m[C]_s}{2V_0} t_x \tag{2.30}$$

This approach has been tested in the determination of iodide as the catalyst of the cerium(IV)–arsenic(III) reaction (17).

## 2.3. CHEMICAL REACTIONS FOR CATALYTIC DETERMINATIONS

In catalytic methods, the main reaction in which the monitored species participates is termed the *indicator reaction*. There are roughly 300 different indicator reactions for determining about 45 elements. The overwhelming majority of indicator reactions for catalytic determinations involve redox systems. Tables 2.2 and 2.3 list typical indicator reaction systems of the redox type and the catalysts determined by them. Other tabulations and, in some cases, detailed information on some of these systems can be found in some reviews and monographs (7, 14, 18–27). Inspection of Tables 2.2 and 2.3 reveals

**Table 2.2. Typical Inorganic Redox Systems Used for Determination of Catalysts**

| Oxidizing Agent | Reducing Agent | Typical Catalysts |
|---|---|---|
| $H_2O_2$ | $S_2O_3^{2-}$ | Ti, Zr, Th, V, Nb, Ta, Mo, W |
| | $I^-$ | Cr, Pb, Ag, Zr, Hf, Th, Ta, Mo, W, Fe |
| | $H_2O_2$ | Cu, Mn, Fe, Pd |
| | $NH_3$ (basic soln) | Cu, Co |
| $ClO_3^-$ | $I^-$ | V, Re, Ru, Os |
| $TeO_4^{2-}$ | $Sn^{2+}$ | Re |
| $MnO_4^-$ | $AsO_2^-$ | Os |
| $Fe^{3+}$ | $S_2O_3^{2-}$ | Cu |
| $Ce^{4+}$ | $Cl^-$ | Ag |
| | $AsO_2^-$ | Os, I, Ru, Hg |
| $Ag^+$ | $Fe^{2+}$ | Au |
| $Ni^{2+}$ | $H_2PO_2^-$ | Ru, Os, Pd, Pt |
| $I_2$ | $N_3^-$ | $S^{2-}$, $S_2O_3^{2-}$, $CNS^-$ |
| $BrO_3^-$ | $I^-$ | V, Os, Mo |
| $IO_4^-$ | $I^-$ | Ru, Ir |
| $BrO^-$ | $Mn^{2+}$ | Rh |
| $S_2O_8^{2-}$ | $I^-$ | Cu, Fe |
| | $Mn^{2+}$ | Ag |

**Table 2.3. Typical Organic Redox Systems Used for Determination of Catalysts**

| Oxidizing Agent | Reducing Agent | Typical Catalyst |
|---|---|---|
| $H_2O_2$ | Aminoazo dyes | Fe, Cu, Cr |
| | Leuco base of variamine blue B | Cu |
| | p-Phenylenediamine | Cu, Fe, Os |
| | p-Phenetidine | Cu, Fe |
| | o-Dianisidine | Cr |
| | o-Tolidine | Cr |
| | Bordeaux S | Co |
| $O_2$ | Leuco base of methylene blue | Cu |
| | Ascorbic acid | Cu |
| $BrO_3^-$ | p-Phenetidine | V, Mo, Os |
| | Azine dyes | Os |
| | $\alpha$-Naphthol | V, Mo |
| $IO_3^-$ | Hydroquinone | Mo |
| $IO_4^-$ | Triphenylmethane dyes | Mn, V |
| | Sulfanilic acid | Mn |
| $Ce^{4+}$ | Diphenylamine | Ir |
| $S_2O_8^{2-}$ | Pyrogallol Blue | Pb |
| | p-Phenetidine | Ag, Mn |
| Methylene blue | $S^{2-}$ | Se |
| $ClO_3^-$ | Phenylhydrazine-p-sulfonic acid | Se |

two dominating features: (a) the catalytic species generally determined are overwhelmingly cations, most of them from transition metal elements, and (b) organic species act mostly as reducing agents in the corresponding indicator reactions. Through the years more catalytic determinations and indicator reactions have been proposed for copper than for any other element. Determinations for osmium and ruthenium have come to be so frequently reported that a review with 46 references has been published on the catalytic determinations of these elements alone (28). Although the number of indicator reactions available in the literature is very extensive, many of them do not offer much advantage over others.

The types of species encountered in indicator reactions can be divided into classes (29): (a) oxidants and reductants participating in reactions through their $d$ orbitals, and (b) oxidants and reductants participating in reactions through their $s$ and $p$ orbitals. Examples of the first class are Fe(III) and $MnO_4^-$ as oxidants and Fe(II), Ti(III), V(III), and $VO_2^{2+}$ as reductants. Examples of the second class are $H_2O_2$, $ClO_3^-$, $S_2O_8^{2-}$, $BrO_3^-$, and $IO_4^-$ as oxidants and $I^-$, $Cl^-$, $N_3^-$, and organic amines and phenols as reductants. Most of the indicator reactions involve the second class of reactants since they are slower than those of the first class and are consequently easier to manipulate under average laboratory conditions. Redox species react fairly fast via $d$ orbitals because they can participate in a lot of overlapping and are reasonably diffuse, in contrast with $s$ and $p$ orbitals, which are of the penetrating type. Interactions between oxidizing agents reacting via $d$ orbitals and reducing agents reacting through $s$ and $p$ orbitals are relatively slow. The $d$ orbital type of oxidizing agent also reacts slowly with some polyatomic species. Reactions between polyatomic $s$ and $p$ orbital oxidants and reductants are very slow, and it is in these that chemists resort to catalysts to increase rates. Tables 2.4 and 2.5 summarize examples of these trends in reactivity. In homogeneous systems, the catalyst can either change oxidation state(s) in its participation in the catalytic cycle or not undergo such change.

As noted earlier, the indicator reaction is always thermodynamically favored, but kinetic complications may seriously hinder or, for all practical

**Table 2.4. Relative Rates of Interaction for Oxidizing Agents Reacting via $d$ Orbitals**

| | Oxidizing Agents | | | | | | |
|---|---|---|---|---|---|---|---|
| | $Mn^{3+}$ | $Fe^{3+}$ | $Cu^{2+}$ | $Ce^{4+}$ | $VOOH^{2+}$ | $CrO_4^{2-}$ | $MnO_4^-$ |
| | *Monoatomic Reductants* | | | | | | |
| $Cr^{2+}$ | Fast | Fast | | | Fast | Fast | |
| $Fe^{2+}$ | Fast | Slow | | Fast | Fast | Fast | Fast |
| $Ti^{3+}$ | Fast | Fast | | | Fast | | |
| $Cu^+$ | | Fast | | | | | |
| $Sn^{2+}$ | Fast | Slow | | Fast | Fast | Fast | Fast |
| $I^-$ | | Slow | Slow | | Average | | Fast |
| | *Polyatomic Reductants* | | | | | | |
| $Hg_2^{2+}$ | | Fast | | Slow | Fast | | Slow |
| $H_2PO_3^-$ | | | | Slow | Slow | | Slow |
| $AsO_2^-$ | Fast | Slow | Fast | Slow | Slow | Slow | Slow |
| $SO_3^{2-}$ | | Slow | | | Slow | Slow | Average |
| $S_2O_3^{2-}$ | | Slow | | Fast | Slow | | Fast |
| $C_2O_4^{2-}$ | Fast | | | Slow | | | Slow |
| Ascorbate | Fast | Fast | | Fast | Fast | Fast | |
| $N_2H_4$ | Fast | | | Slow | Average | | |
| ArOH | Fast | | | | Slow | | |
| $ArNH_2$ | Fast | | | | Slow | | |

*Source*: Ref. 29, reproduced with permission.

**Table 2.5. Relative Rates of Oxidation of $s$ and $p$ Orbital Type of Reducing Agents and $Fe^{2+}$ by $s$ and $p$ Orbital Type of Polyatomic Species**

| Oxidizing Agent | Reducing Agent | | | | |
|---|---|---|---|---|---|
| | $Fe^{2+}$ | $Sn^{2+}$ | $I^-$ | $AsO_2^-$ | Ascorbate |
| $ClO_3^-$ | Slow | Slow | Slow | Slow | Slow |
| $BrO_3^-$ | | | Slow | Slow | |
| $IO_3^-$ | | | | Slow | Slow |
| $ClO_4^-$ | | Slow | Slow | Slow | Slow |
| $H_2O_2$ | | | Slow | Slow | |
| $S_2O_8^{2-}$ | | | Slow | Slow | |

*Source*: Ref. 29, reproduced with permission.

purposes, prevent the reaction within the time scale of experimental observation. The catalyst then facilitates the process of reactants going to products in the indicator reaction. Mechanistically this facilitation may be quite complex, as reflected in the form of experimentally derived rate expressions. An example of this is the "simplified" rate equation derived by Rodriguez and Pardue (30) for the catalytic effect of iodide ions on the Ce(IV)–As(III) reaction:

$$Rate = \frac{(5.82 \times 10^3)[Ce(IV)][As(III)](8.46[Ce(IV)] + (3.60 \times 10^5)[As(III)] + 8.46}{(1.75 \times 10^3)[As(III)]^2 + (1 + [Ce(IV)])(21.5[As(III)] + 5.11[Ce(IV)])} [I]_{tot}$$

This equation is indicative of the complex dependence of the rate on the reactants in the indicator reaction. Some mechanistic observations can, however, be pointed out:

1. For catalysts changing oxidation state, the interaction between the catalyst and one of the reactants (e.g., organic amines and phenols) involves substitution of water molecules in the catalyst coordination sphere by the reactant acting as ligand. This results in charge transfer complex formation (31).

2. In some reactions, transfer of more than one electron between reactants is facilitated by catalysts capable of existing in at least two oxidation states under the given experimental conditions [e.g., oxidation of oxalic acid by permanganate catalyzed by Mn(II) ions, which occurs by successive reactions with catalyst participation (32, 33)].

3. In some cases where more than one electron is transferred between reactants, the formation of unstable intermediates, involving the catalyst, facilitates one-electron transfer steps [e.g., the iodine–iodide action in the Ce(IV)–As(III) reaction (30)].

4. In reactions in which the oxidation state of the catalyst remains unchanged, two general groups can be distinguished: (a) reactions in which $H_2O_2$ acts as oxidant and catalytic cations (with high charges) form unstable peroxocomplexes decomposing to yield highly reactive free radicals (e.g., OH·, OH⁻ or $O_2H$· radicals), and (b) catalytic decomposition, hydrolysis, or substitution reactions.

A generalized view along these lines, on the mechanisms of catalytic reactions of analytical interest, can be found in the first part of a review by Bontchev (34).

### 2.4. LIMITS OF DETECTION, SENSITIVITY, SELECTIVITY, AND OPTIMIZATION IN CATALYTIC DETERMINATIONS

#### 2.4.1. Limits of Detection and Sensitivity

Low *limits of detection* (lowest concentration of catalyst detected with high probability) and high *sensitivities* (represented by the slope of calibration graphs) are recognized as the major advantages of catalytic determinations. These two concepts are of such importance in catalytic methods that a summary of their basic mathematical expressions is given in Table 2.6. These expressions show that the initial concentration of monitored species, $[R]_0$, is critical when a derivative approach is used in initial rate determinations. If the variable-time procedure is used, two factors exert opposite effects on both detectability and sensitivity. These factors are (a) the minimum detectable change in analytical signal, and (b) the maximum time interval with which it would be reasonable to work. In fixed-time procedures time is the critical variable. In all these cases, however, $k_c$ must be maximized and $k_u$ minimized since the ratio $k_u/k_c$ dictates the limits of detection.

Unfortunately the literature many times does not make a clear distinction between detectability and sensitivity. From a statistical viewpoint the limit of detection can be calculated from the expression

$$LD = \chi_b + 3s_b$$

where $\chi_b$ = average of blank readings, and $s_b$ = sample standard deviation for the collection of blank readings. In practice, of course, LD must be numerically translated to the concentration units of the sought-for species.

**Table 2.6. Linear Relationships between Sought-for Species and Kinetic Quantities in Common Cases of Catalytically Based Determinations**[a]

| Method | Detectability | Sensitivity |
|--------|---------------|-------------|
| Initial rate [derivative] | $[C]_0 = -\dfrac{d[R]}{dt}\dfrac{1}{k_c[R]_0} - \dfrac{k_u}{k_c}$ | $\dfrac{\Delta(d[R]/dt)}{\Delta[C]_0} = k_c[R]_0$ |
| Variable time[b] [integral][c] | $[C]_0 = \dfrac{K}{\Delta t\, k_c} - \dfrac{k_u}{k_c}$ | $\dfrac{\Delta(1/\Delta t)}{\Delta[C]_0} = \dfrac{k_c}{K}$ |
| Fixed time [integral][c] | $[C]_0 = -\dfrac{\ln[R]_1}{k_c t_1} + \dfrac{\ln[R]_0}{k_c t_1} - \dfrac{k_u}{k_c}$ | $\dfrac{\Delta(\ln[R]_1)}{\Delta[C]_0} = k_c t_1$ |

[a] These expressions are restricted to catalyst determination under conditions in which eq. 2.21 is valid.
[b] $K = -[\ln([R]_2/[R]_1)] = $ constant.
[c] The term *integral* refers to the mathematical treatment of eq. 2.21. The same expressions are valid in case of initial-rate estimation by the fixed- or variable-time procedure.
*Source*: Reprinted from Ref. 22 with permission. Copyright CRC Press, Inc., Boca Raton, FL.

The limit of detection should not be confused with the *limit of determination*, which is calculated using

$$LD_m = \chi_b + 10s_b$$

A practical discussion of these concepts has been prepared and published jointly by a committee and a subcommittee of the American Chemical Society concerned with environmental problems (35).

Yatsimirskii, the author of the first monograph on kinetic methods (7), has indicated in the introduction to his book that

Kinetic methods of analysis using catalytic reactions are of extremely high sensitivity. They are equal to activation analysis methods and superior to spectral and spectrophotometric methods.

The word *sensitivity* in this statement was probably meant conceptually to equal limit of detection. A numerical exercise backing up Yatsimirskii's assertion follows. Starting with eqs. 2.13 and 2.20 we may write

$$-\frac{d[R]}{dt} = k[C]_0$$

since from a practical viewpoint $k_u \ll k_c$. Infinitely small values can be replaced by finite increments such as

$$[C]_0 = \frac{\Delta[R]}{\Delta t} \cdot \frac{1}{k_c} \tag{2.31}$$

If the change in [R] is followed spectrophotometrically (as is commonly done), we can assume conservatively that a change in $\Delta[R]$ equal to about $1 \times 10^{-6} M$ can be detected in initial rate measurements of a catalyzed chemical reaction with a rate coefficient of $k_c = 10^4$ min. Arbitrarily we can assume a $\Delta t$ of 1 min; this is conservative again, since larger values of $\Delta t$ are recommended in the literature. From eq. 2.31 we can then conclude that under these conditions $[C]_{0 \min} = 10^{-10} M$.

Consider now the same determination of $[C]_0$ by a conventional spectrophotometric method (e.g., if C is a metal ion that is complexed by an organic ligand to form a chromophore) and a maximum tolerable relative concentration error of 5%. In the 80-to-100% transmittance range (36) the relative concentration error can be described as

$$\left(\frac{\Delta[C]_0}{[C]_0}\right) = \frac{30}{100 - \%T} \tag{2.32}$$

where $\%T$, measuring for 5% error, would be equal to 94.0. This value corresponds to an absorbance of 0.027 units. According to Lambert–Beer's law, the chromophore should have a molar absorptivity $\varepsilon$ of about $3 \times 10^8 M^{-1} \cdot cm^{-1}$ in order for the determination of $[C]_0 = 10^{-10} M$ to be possible. This value of $\varepsilon$ is at least three orders of magnitude larger than the maximum theoretical value of $10^5 M^{-1} \cdot cm^{-1}$ (37). This allows the conclusion that for this particular case, using the same instrumental monitoring, the catalytic determination offers considerably lower limits of detection than determination by equilibrium spectrophotometry.

### 2.4.2. Selectivity and Specificity

If limits of detection and sensitivity can be singled out as advantages in catalytic determinations, selectivity, on the other hand, can be considered to limit the practical applications of these determinations (38). According to IUPAC recommendations (39), *selectivity* denotes the extent to which other chemical species interfere with the determination of a given species by a given method. Specificity and selectivity, like limits of detection and sensitivity, are many times misused in the literature. *Specificity* is the case in which no interferent effects are known (39). In actuality, when catalytic methods are considered, specificity is a rarity outside the use of enzyme-catalyzed reactions. A few exceptions can be found, such as the dimerization of quinoline aldehyde, which seems to be specifically catalyzed by cyanide ions (40); commonly, however, nonenzymatic catalytic methods offer poor selectivity. This is because catalytic behavior is dictated by size, structure, charge, and coordination sphere of the catalyst, which results in equivalent catalytic effects by similar chemical species.

Attempts to give quantitative meaning to selectivity and specificity can be found in the literature (38, 41). These attempts consider that selectivity applies to the system as a whole, and as such what is needed are calibration functions represented by rate equations for different conditions. These rate equations must describe the concerted action of the different species involved. These expressions can be written as follows (38):

$$r_1 - r_{10} = k_{11}\pi_{11}[C_1] + k_{12}\pi_{12}[C]_2 + \cdots + k_{1n}\pi_{1n}[C]_n$$

$$r_2 - r_{20} = k_{21}\pi_{21}[C_1] + k_{22}\pi_{22}[C]_2 + \cdots + k_{2n}\pi_{2n}[C_n]$$

$$\vdots$$

$$r_m - r_{m0} = k_{m1}\pi_{m1}[C_1] + k_{m2}\pi_{m2}[C_2] + \cdots + k_{mn}\pi_{mn}[C_n]$$

with $r_m$ = rates of catalyzed reactions, $r_{m0}$ = corresponding uncatalyzed rates, $k_{ij}$ = rate coefficients, $\pi'_{ij}$ = concentration products of reactants in the indicator reaction according to the reaction order, and [C] = concentration of the corresponding catalyst. Quantitative expressions for the selectivity $\Xi$ and specificity $\Psi$ can be written with the help of a matrix containing all partial sensitivities, as functions of the slopes of calibration functions, $\gamma$:

$$\Xi = \underset{i=1 \cdots n}{\text{Min}} \left( \frac{\gamma_{ii}}{\sum \gamma_{ik} - \gamma_{ii}} - 1 \right) \tag{2.33}$$

$$\Psi = \frac{\gamma_{aa}}{\sum \gamma_{kk} - \gamma_{aa}} - 1 \tag{2.34}$$

High selectivity is obtained if, in the matrix represented by eq. 2.33, all the elements not just those of the main diagonal, approach zero. According to eq. 2.34, specificity is realized if the element $\gamma_{aa}$ of the matrix is different from zero. In absence of synergistic effects between catalysts, these formulas work satisfactorily.

From a practical viewpoint, several routes are opened to improving selectivity in catalytic methods. Table 2.7 gives a sample of those avenues.

**Table 2.7. Selected Examples of Different Approaches to Improve Selectivity in Catalytic Determinations**

| System | Approach used to improve selectivity | Reference |
|---|---|---|
| $I^- + H_2O_2$ (catalyzed by Zr and Hf) | *Variation of pH.* Zr shows highest activity at pH 1.1 and Hf at pH 2.2 owing to differences in stabilities of their hydroxo complexes. Selective determination of Zr and Hf is accompanied by pH selection. | 42 |
| $Ce(IV) + As(III)$ (catalyzed by Ru and Os) | *Variation of reagent concentration in the indicator reaction.* Different ratios of $[As(III)]/[Ce(IV)]$ allow selective determination of Ru and Os | 43 |
| 1-Naphthylamine + $NO_3^-$ (determination of Ru and Os). | *Inhibition and activation.* Ru is determined in presence of 1,10-phenanthroline, which by complexation inhibits Os catalysis. The Os catalysis, on the other hand, is activated by 8-hydroxyquinoline, which improves the selectivity for Os determination. | 44 |
| $I^- + BrO_3^-$ [determination of Mo(VI), W(VI), and Cr(VI)] | *Temperature.* In the temperature range of 0–40°C each catalyst exerts a different extent of catalytic action (see Figure 2.5). | 45 |
| Bromopyrogallol red + $S_2O_8^{2-}$, with 1,10-phenanthroline as activator (determination of Ag) | *Extraction of species of interest into an organic solvent and direct catalytic determination in the organic phase.* Determination in a mixture of nitrobenzene, dioxane, and water, after extraction of silver with 1,10-phenanthroline into nitrobenzene in presence of bromopyrogallol red. | 46 |

## 2.4.3. Optimization of Catalytic Methods

Commonly the optimization of catalytic methods is accomplished by sequentially optimizing one variable at a time, keeping all other variables constant. The introduction of microcomputers into the analytical laboratory has made available relatively sophisticated mathematical tools of data evaluation and method optimization. As a result, numerical models and sequential simplex methodology have been explored for the optimization of catalytic determinations (47–49).

The copper-catalyzed decomposition (by dissolved oxygen) of hydrogen peroxide in the presence of pyridine as activator, a relatively simple situation, has been successfully modeled by a numerical approach (49). However, this approach to modeling requires knowledge of equilibria and kinetics of the reaction under study. Response surfaces and both fixed-size and variable-size simplex (50) have also been used with the same system to obtain optimal concentrations for hydrogen peroxide, pyridine, and hydrogen ion (49). Simplex has also been used to maximize the sensitivity of the catalytic polarographic wave of the uranyl nitrate system for uranium determination (49).

## 2.5. SOME APPLICATIONS OF NONENZYMATIC CATALYTIC DETERMINATIONS

The number of methods recorded in the literature for catalytic determinations other than enzymatic ones is very large as a result of the extensive availability of indicator reactions to serve such purposes. Early pertinent references and practical details can be found in Yatsimirskii's monograph (7). Recent reviews in the "fundamental reviews" series of *Analytical Chemistry* (24–27) update the records of this type of application.

### 2.5.1. The Sandell–Kolthoff Reaction

Historically and practically the As(III)–Ce(IV) indicator reaction (the *Sandell–Kolthoff reaction*), which is catalyzed by iodide and some iodine-containing compounds, has commanded a sustained attention in the development of catalytic methods. This indicator reaction has become the standard for the determination of iodide (and iodine and iodate) at very low concentrations in a large variety of samples, such as table salt, water (sea, potable, and natural), grass and vegetables, filter paper, and biological fluids (e.g., serum and urine). It has been found particularly useful in the determination of protein-bound iodine (PBI) because there are no common species in ashed serum that interfere either by altering the catalytic cycle or by reaction with Ce(IV) or As(III). The relatively high chloride content of serum samples is beneficial because chloride ion apparently prevents a side reaction (51). The procedure calls for precipitation and separation (by filtration) of proteins that are then dry-ashed in a furnace. The ash is dissolved in a mixture of sulfuric and hydrochloric acid, and the reaction is started by addition of Ce(IV) and

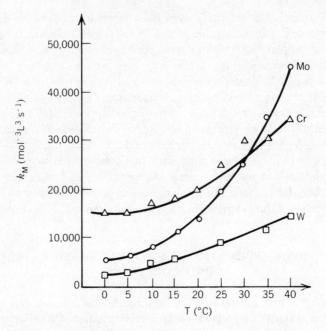

**Figure 2.5.** Variation of rate coefficient with temperature for molybdenum, chromium, and tungsten as catalysts of the iodide–bromate indicator reaction. [Reprinted with permission from Ref. 45.]

As(III) reagent solutions. Measurement is commonly done by photometric monitoring of the Ce(IV) concentration at 420 nm.

The catalytic action of iodide has also been used in almost all conceivable ways for the indirect determination of a large number of chemical species. In their studies on the chemistry of this chemical system Sandell and Kolthoff (52) noted the inhibitory effects of fluoride [which complexes Ce(IV)], mercury(II) and silver(I) (which complex and precipitate iodide) and cyanide. Since then a great variety of methods have been reported for the indirect determination of silver and mercury based on their inhibitory effect; recent applications of this effect can be found in the literature (53, 54), particularly some involving catalytic end-point indications (55), a topic discussed in Chapter 4. Several pharmaceutical products containing mercury compounds as active species have also been analyzed by use of the Ce(IV)–As(III) reaction with iodide as the catalytic titrant (56); liquid-phase oxidation of the sample was used to free the mercury for titration.

A detailed study aimed at developing optimal conditions for the determination of iodide using the Ce(IV)–As(III) indicator reaction in an air-segmented continuous-flow system has been provided by Keller et al. (57), Trimarchi and co-workers discussed the advantages of automated determination of PBI in serum, again using air-segmented continuous-flow sample processing and the same chemistry for determination (58).

The application of this indicator reaction for the determination of iodine-containing compounds is of unique interest. For instance, to determine thyroid hormones Leopold and Knapp used nitric acid in the medium (59) because they found the iodide catalysis 20 times better in such a medium than in the commonly used sulfuric acid. Nachtmann et al. (60) have proposed a post-column catalytic detection for high-performance liquid chromatography of iodine-substituted molecules (e.g., hormones of the thyroid gland). They report limits of detection for tetraiodothyronine and triiodothyronine at the subnanogram level, sufficient for their determination in human plasma. Ultraviolet- and fluorescence-based detectors have been found unsatisfactory for the same purposes (60). Figure 2.6 illustrates the type of system used for

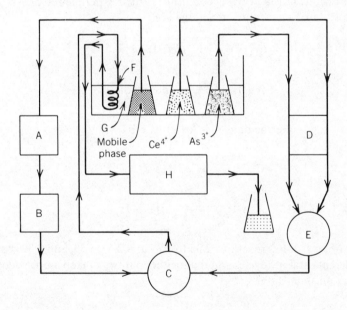

**Figure 2.6.** Post-column catalytic determination of iodine-containing species: A, pump for mobile phase eluted from the liquid chromatographic column; B, injection system; C, mixing device; D, pump for reagent solutions; E, mixing device; F, reaction capillary; G, thermostat; H, spectrophotometer. [Reprinted with permission from Ref. 60.]

post-column catalytic detection proposed by Nachtmann et al. (60).

Elverog and Carr (61) employed the Ce(IV)–As(III) reaction in an un-segmented continuous-flow thermochemical reactor–detector system yielding a minimum detectable quantity for iodide of 0.15 ng, perhaps the lowest reported for a thermochemical detection.

Several kinetic studies have resulted in postulated mechanisms for the Ce(IV)–As(III) reaction. Because all reactions involving iodine are known to be rapid, the pronounced catalytic effect of iodine or iodide has been postulated to occur through the following sequence (62):

$$Ce(IV) + I^- \rightarrow I^\circ + Ce(III)$$

$$2I^\circ \rightarrow I_2$$

$$I_2 + H_2O \leftrightharpoons HIO + H^+ + I^-$$

$$H_3AsO_3 + HIO \rightarrow H_3AsO_4 + H^+ + I^-$$

As a result of thoroughly investigating the reaction in a wide range of conditions, Rodriguez and Pardue (30) have postulated pathways that yield rate expressions satisfactorily accounting for their experimental observations. Their postulated elementary reactions are.

$$I^- + Ce(IV) \xrightarrow{k_1} Ce(III) + I^\circ \tag{2.35}$$

$$I^\circ + As(III) \underset{k_{-2}}{\overset{k_2}{\rightleftharpoons}} \text{intermediate} \tag{2.36}$$

$$\text{intermediate} \xrightarrow{k_3} As(V) + Ce(III) + I^- \tag{2.37}$$

$$I^\circ + Ce(IV) \xrightarrow{k_4} Ce(III) + I^+ \tag{2.38}$$

$$I^+ + As(III) \underset{k_{-5}}{\overset{k_5}{\rightleftharpoons}} As(V) + I^- \tag{2.39}$$

They observed that reactions 2.35 to 2.37 and 2.35, 2.38, and 2.39 represent two different catalytic cycles and that neither pathway taken alone adequately describes all their data.

### 2.5.2. Miscellaneous Applications of Catalyzed Reactions

The low limits of detection of catalytic methods have been exploited by Yatsimirskii et al. (63) for the detection of gas or liquid leaks from pressure vessels. A catalyst is contained in the pressurized gas or liquid while an aqueous mixture of the indicator reaction is applied to the external surface of the vessel; the detection of the leak is based on a color change as the result of the catalytic effect. They tested two systems: (a) the oxidation of $CNS^-$ by Fe(III) in the presence of $NO_2^-$, in which the red iron–thiocyanate complex(es) are destroyed by the catalytic action of iodine; and (b) the oxidation of o-toluidine by chlorate ion, in which osmium as the catalyst causes the formation of a blue color.

Lankmayr et al. (64, 65) exploited the catalytic effect of iodide on the reaction of $N,N'$-tetramethyldiaminodiphenylmethane with chloramine T for liquid chromatographic post-column detection of enantiomeric iodinated thyronines in blood serum. The color produced in the reaction is monitored at 600 nm. The limit of detection for isomeric tetraiodothyronines in serum is reported to be about 3 nmol/L.

Sample preparation for the determination of vanadium in petroleum oils is generally performed by dry or wet ashing, which has the possibility of losing metallic species through volatlllization and introducing contaminants. The oxygen flask combustion method for sample preparation avoids these problems, but it requires a small sample size (less than 100 mg) and is thus incompatible with the lack of sensitivity of conventional determination procedures. Catalytic determination of vanadium using the gallic acid–bromate indicator reaction provides sufficiently low limits of detection and satisfactory sensitivity even in small samples and has been coupled with the oxygen flask combustion method for the determination of vanadium in crude oil and residual fuel oils (66).

Catalytic reactions are commonly used for the determination of catalysts and only seldom for reactants. Grases and March (67) reported the determination of technetium(VII) by means of the copper-catalyzed oxidation of variamine blue by Tc(VII). The concentration of Tc(VII) amenable to determination is 0.2 to 2.4 $\mu$g/mL. Using the method of tangents or a fixed-time procedure, the authors applied the method to the analysis of synthetic nuclear fuels.

The use of unsegmented continuous-flow sample–reagent(s) processing greatly facilitates the incorporation of ancillary procedures (68) such as on-line ion-exchange separations. This permits one to improve selectivity in catalytic determinations that as already noted are inherently unselective except for those involving enzyme catalysis. As an example, the method proposed by Yamane for manganese determination in natural waters can be cited (69). A continuous-flow system with on-line cation-exchange separation and catalytic detection afforded a limit of detection of 0.2 ng/mL with competitive selectivity for manganese. The chemistry used for catalytic detection was based on the hydrogen peroxide oxidation of protocatechuic acid in basic medium.

Photometric detection (at 480 nm) of a reaction product provided the basis for determination.

## REFERENCES

1. S. Glasstone, *Textbook of Physical Chemistry*, Van Nostrand, New York, 1940, p. 1103.
2. W. Ostwald, *Phys. Z.*, **3**, 313 (1902).
3. G. A. Vaughan and J. J. Swithenbank, *Analyst*, **90**, 594 (1965).
4. V. J. Vijgand and F. F. Gaal, *Talanta*, **14**, 345 (1967).
5. V. J. Vijgand, T. A. Kiss, F. F. Gaal, and I. J. Zsigrai, *Talanta*, **15**, 699 (1968).
6. V. J. Vijgand, *Talanta*, **17**, 415 (1970).
7. K. B. Yatsimirskii, *Kinetic Methods of Analysis*, Pergamon, Oxford, 1966, p. 60.
8. K. B. Yatsimirskii, in *MTP International Review of Science: Physical Chemistry. Series One, Part* 12, T. S. West, Ed., Butterworths–University Park, Oxford, 1973, Chapter 5, p. 206.
9. C. W. Pyun, *J. Chem. Educ.*, **48**, 194 (1971).
10. M. Otto, G. Schobel, and G. Werner, *Anal. Chim. Acta*, **147**, 287 (1983).
11. H. A. Mottola, *MPI Appl. Notes*, **6**, 17 (1971).
12. J. D. Ingle, Jr., and S. R. Crouch, *Anal. Chem.*, **43**, 697 (1971).
13. F. J. Holler, R. K. Calhoun, and S. F. McClanahan, *Anal. Chem.*, **54**, 755 (1982).
14. M. Kopanica and V. Stara, in *Wilson and Wilson's Comprehensive Analytical Chemistry*, G. Svehla, Ed., Elsevier, Amsterdam, 1983, Vol. 28, p. 122.
15. H. Landolt, *Ber. Dtsch. Chem. Ges.*, **19**, 1317 (1886).
16. G. Svehla, *Analyst (London)*, **94**, 513 (1969).
17. G. Lopez-Cueto and A. F. Cueto-Rejon, *Anal. Chem.*, **59**, 645 (1987).
18. H. Muller, M. Otto, and G. Werner, *Katalytische Methoden in der Spurenanalyse*, Akademische Verlagsgesellschaft Geets & Portig K.-G., Leipzig, 1980.
19. Z. Gregorowicz and T. Suwinska, *Chem. Anal. (Warsaw)*, **11**, 3 (1966).
20. A.-M. Gary and J.-P. Schwing, *Bull. Soc. Chim. Fr.*, 3657 (1972).
21. H. V. Malmstadt, C. J. Delaney, and E. A. Cordos, *CRC Crit. Rev. Anal. Chem.*, **2**, 559 (1972).
22. H. A. Mottola, *CRC Crit. Rev. Anal. Chem.*, **4**, 229 (1975).
23. H. Muller, *CRC Crit. Rev. Anal. Chem.*, **13**, 313 (1982).
24. H. A. Mottola and H. B. Mark, Jr., *Anal. Chem.*, **52**, 31R (1980).
25. H. A. Mottola and H. B. Mark, Jr., *Anal. Chem.*, **54**, 62R (1982).
26. H. A. Mottola and H. B. Mark, Jr., *Anal. Chem.*, **56**, 96R (1984).
27. H. A. Mottola and H. B. Mark, Jr., *Anal. Chem.*, **58**, 264R (1986).
28. V. I. Shlenskaya, V. P. Khvostova, and G. I. Kadyrova, *Zh. Anal. Khim.*, **28**, 779 (1973); *Chem. Abstr.*, **79**, 26724v (1973).

29. K. B. Yatsimirskii, *Pure Appl. Chem.*, **27**, 251 (1971).

30. P. A. Rodriguez and H. L. Pardue, *Anal. Chem.*, **41**, 1369 (1969).

31. H. Muller and M. Otto, *Z. Chem.*, **14**, 159 (1974).

32. H. Taube, *J. Am. Chem. Soc.*, **70**, 1216 (1948).

33. S. J. Adler and R. M. Noyes, *J. Am. Chem. Soc.*, **77**, 2036 (1955).

34. P. R. Bontchev, *Talanta*, **17**, 499 (1970).

35. ACS Committee on Environmental Improvement and Subcommittee on Environmental Analytical Chemistry, *Anal. Chem.*, **52**, 2242 (1980).

36. E. B. Sandell, *Colorimetric Determination of Traces of Metals*, Wiley-Interscience, New York, 1959, pp. 97–98.

37. E. A. Baude, *J. Chem. Soc.*, 379 (1950).

38. G. Werner, *Quim. Anal. (Barcelona)*, II Extra, 68 (1983).

39. G. Den Boef and A. Hulanicki, *Pure Appl. Chem.*, **55**, 553 (1983).

40. G. Werner, M. Hanrieder, and H. Muller, *Proc. 8th Symp. Microchem.*, Graz, Austria, 1980, p. 456.

41. H. Kaiser, *Fresenius' Z. Anal. Chem.*, **260**, 252 (1972).

42. K. B. Yatsimirskii and L. P. Raizman, *Zh. Anal. Khim.*, **18**, 29 (1963).

43. L. B. Worthington and H. L. Pardue, *Anal. Chem.*, **42**, 1157 (1970).

44. H. Muller and G. Werner, *Z. Chem.*, **14**, 159 (1974).

45. C. M. Wolff and J. P. Schwing, *Bull. Soc. Chim. Fr.*, 679 (1976).

46. H. Muller, H. Schurig, and G. Werner, *Talanta*, **21**, 581 (1974).

47. A. MacDonald, K. W. Chan, and T. A. Nieman, *Anal. Chem.*, **51**, 2077 (1979).

48. M. Otto and G. Werner, *Anal. Chim. Acta*, **128**, 177 (1981).

49. A. P. Wade, *Anal. Proc. (London)*, **20**, 108 (1983).

50. S. L. Morgan and S. N. Deming, *Anal. Chem.*, **46**, 1170 (1974).

51. M. Dubravcic, *Analyst (London)*, **80**, 146, 295 (1955).

52. E. B. Sandell and I. M. Kolthoff, *Mikrochim. Acta*, p. 9 (1937).

53. P. A. Rodriguez and H. L. Pardue, *Anal. Chem.*, **41**, 1376 (1969).

54. P. J. Ke and R. J. Thibert, *Mikrochim. Acta*, p. 768 (1972).

55. F. F. Gaal and B. F. Abramovic, *Talanta*, **27**, 733 (1980).

56. F. F. Gaal and B. F. Abramovic, *Mikrochim. Acta*, p. 465 (1982).

57. H. E. Keller, D. Doenecke, K. Weidler, and W. Leppla, *Ann. N.Y. Acad. Sci.*, **220**, 1 (1973).

58. I. Trimarchi, F. Munao, G. Mobilia, and M. L. Calisto, *Quad. Sclavo Diagn. Clin. Lab.*, **8**, 1022 (1972).

59. G. Knapp and H. Leopold, *Anal. Chem.*, **46**, 719 (1974).

60. F. Nachtmann, G. Knapp, and H. Spitzy, *J. Chromatogr.*, **149**, 693 (1978).

61. J. M. Elverog and P. W. Carr, *Anal. Chim. Acta*, **121**, 135 (1980).

62. W. C. Bray, *Chem. Rev.*, **10**, 161 (1932).

63. K. B. Yatsimirskii, A. I. Zapunnyi, L. I. Budarin, L. S. Fel'dman, and Kazakevich, *Zavod. Lab.*, **43**, 920 (1977); *Chem. Abstr.*, **88**, 138, 304a (1978).

64. E. P. Lankmayr, B. Maichin, and G. Knapp, *Fresenius' Z. Anal. Chem.*, **301**, 187 (1980).

65. E. P. Lankmayr, B. Maichin, and G. Knapp, *J. Chromatogr.*, **224**, 239 (1981).

66. T. Fukasawa and T. Yamane, *Anal. Chim. Acta*, **113**, 123 (1980).

67. F. Grases and J. G. March, *Analyst (London)*, **110**, 975 (1985).

68. H. A. Mottola, *Anal. Chim. Acta*, **180**, 26 (1986).

69. T. Yamane, *Anal. Sci.*, **2**, 191 (1986).

CHAPTER

3

# CATALYTIC METHODS:
# HOMOGENEOUS ENZYMATIC METHODS

Enzyme-catalyzed reactions are used analytically to determine enzyme activities and substrate concentrations, both parameters of importance in clinical diagnoses. The use of enzymes in the diagnosis of disease is not only an important benefit derived from biochemical research that has intensified since the 1940s; it has also provided the avenue that introduced clinical analysis as a significant branch of analytical chemistry.

The large number of enzymatic determinations performed daily in clinical laboratories around the world is responsible for a fact that comes as a surprise at first: determinations carried out using kinetic-based methods outnumber those carried out by equilibrium-based methods or direct instrumental measurement.

## 3.1.  ENZYMES AS CATALYSTS

The biochemical process known as metabolism is characterized by chemical and physical changes continuously going on in what we call "life." The majority of the biochemical reactions involved in these changes are too slow for life to keep going if enzymes are not present. The fact that enzymes act as catalysts in key biochemical reactions makes life possible. The oxidation of a fatty acid to carbon dioxide and water requires, in the laboratory, rather extreme conditions of pH and temperature as well as corrosive oxidizing acidic mixtures. The same chemistry, however, can occur in living cells with enzymatic help in a gentle yet rapid manner within a narrow pH and temperature range.

In 1835, Jon Jakob Berzelius, who performed some of the earliest studies with enzymes, recognized their chemical actions as catalytic (1). The first enzyme isolated in pure crystalline form was urease from the jack bean. This earned James B. Summer a share of a Nobel Prize in 1947. The requirements of purity, stability, and commercial availability have delayed the application of enzymes as analytical reagents. As stated in Chapter 1, in 1951 Stetter described the use of impure enzyme preparations in a variety of analytical

53

procedures (2), but the real impact of enzymes in chemical analysis was not felt until the mid 1960s and early 1970s (3).

All known enzymes are proteinaceous materials of high molecular weight (10,000 to 2,000,000) made up primarily of chains of amino acids linked together by peptide bonds. Many enzymes need the presence of other species, known as *cofactors*, to act as catalysts. The enzyme–cofactor complex is called an *holoenzyme*; the proteinaceous portion is known as an *apoenzyme*:

$$holoenzyme = apoenzyme + cofactor$$

Apoenzymes may form complexes with different types of cofactors such as: (a) a *coenzyme* (a nonproteinaceous organic species loosely attached to the apoenzyme), (b) a *prosthetic group* (an organic species firmly attached to the apoenzyme), and (b) a *metallic ion*.

The one of enzymes property, that singles them out as unique analytical reagents is their high degree of selectivity or specificity. Some enzymes catalyze only one specific reaction. Other enzymes exhibit catalytic activity on reactions involving only chemical species that have a given functional group (e.g., amino, phosphate, or methyl groups) and are considered to have *group specificity*. *Linkage specificity* is exhibited by enzymes that act only on a certain type of chemical bond, and *stereochemical specificity* by enzymes that exert catalytic action on particular steric or chiral isomers. The high degree of selectivity of enzymes is not enjoyed by cofactors, which may interact with different apoenzymes. Nicotinamide adenine dinucleotide (NAD), for instance, is a cofactor (coenzyme) for several dehydrogenases (e.g., alcohol dehydrogenase, malate dehydrogenase, and lactate dehydrogenase) and acts as a hydrogen acceptor.

A classification of enzymes with respect to the type of chemical reaction they catalyze is given in Table 3.1.

## 3.2. KINETICS OF ENZYME-CATALYZED REACTIONS

A general, simplified mechanism for enzyme-catalyzed reactions can be illustrated as follows:

$$E + S \underset{k_{-1}}{\overset{k_1}{\rightleftharpoons}} ES \underset{k_{-2}}{\overset{k_2}{\rightleftharpoons}} P + E \tag{3.1}$$

where $E$ = enzyme, $S$ = substrate, $ES$ = addition complex, and $P$ = product(s).

Mechanistically the scheme represented by eq. 3.1 has been treated both as a

**Table 3.1. Classification of Enzymes According to Type of Chemical Reaction Catalyzed**

| Chemical Change | Enzymes |
| --- | --- |
| Addition or removal of water | a. *Hydrolases* (e.g., esterases, carbohydrases, nucleases, deaminases, amidases, proteases). |
| | b. *Hydrases* (e.g., fumarase, enolase, carbonic anhydrase). |
| Electron exchange | a. *Oxidases* (e.g., glucose oxidase, galactose oxidase, uricase, amino acid oxidases). |
| | b. *Dehydrogenases* (e.g., alcohol dehydrogenase, hydroxysteroid dehydrogenase, lactate dehydrogenase). |
| Radical transfer | a. *Transglycosidases* (of monosaccharides). |
| | b. *Transphosphorylases and phosphomutases* (of a phosphate group). |
| | c. *Transaminases* (of amino groups). |
| | d. *Transmethylases* (of a methyl group). |
| | e. *Transacetylases* (of an acetyl group). |
| Breaking or forming of a C–C bond | *Desmolases* |

*Source*: From Ref. 4.

pre-equilibrium case (5) and as a steady-state case (6). What follows is a generalized derivation leading to the basic equation providing mathematical justification for substrate and enzyme activity determinations. It is fundamentally based on the Briggs and Haldane scheme (6). The overall rate of substrate–enzyme complex formation can be written as

$$\frac{d[ES]}{dt} = k_1[E][S] + k_{-2}[P][E] - k_{-1}[ES] - k_2[ES] = 0 \qquad (3.2)$$

If initial rate conditions are assumed, $[P] = 0$ for all practical purposes and

$$k_1[E][S] = k_{-1}[ES] + k_2[ES] = [ES](k_{-1} + k_2) \qquad (3.3)$$

which can be rewritten as

$$\frac{[ES]}{[E][S]} = \frac{k_1}{k_{-1} + k_2} = \frac{1}{K_M} \qquad (3.4)$$

where $K_M$ = the Michaelis–Menten constant.

Consider the maximum initial rate $(IR)_{max}$ when all the enzyme is in the complex form ES. Then

$$(IR)_{max} = k_2 [E]_{tot} \qquad (3.5)$$

On the other hand, under any other conditions characterized by $[E]_0 \neq [ES]$,

$$IR = k_2 [ES] \qquad (3.6)$$

Moreover,

$$[E]_{tot} = [E] + [ES]$$

$$[E] = [E]_{tot} - [ES]$$

and

$$[E] = \frac{(IR)_{max}}{k_2} - \frac{IR}{k_2} \qquad (3.7)$$

Substituting in eq. 3.4 gives

$$\frac{1}{K_M} = \frac{k_2 [ES]}{[S]\{(IR)_{max} - IR\}} \qquad (3.8)$$

and, because of eq. 3.6, we can write

$$K_M = [S] \ \frac{(IR)_{max} - IR}{IR} \qquad (3.9)$$

Rearranging and simplifying gives

$$(IR) = \frac{[S] (IR)_{max}}{K_M + [S]} \qquad (3.10)$$

Considering eq. 3.5, we finally obtain

$$IR = -\frac{d[S]}{dt} = \frac{k_2 [S] [E]_0}{K_M + [S]} \qquad (3.11)$$

The relationship given by eq. 3.11 is the basic one to guide analytical determinations of enzyme concentration (activity) as well as substrate concentration. It indicates that, experimentally, the initial rate is directly proportional to enzyme concentration when $[S] \gg K_M$, since then the following applies:

$$IR = (IR)_{max} = k_2[E]_0 \qquad (3.12)$$

Moreover, when $[S] \ll K_M$, the initial rate is directly proportional to substrate concentration:

$$IR = \frac{k_2[E]_0}{K_M}[S] = \frac{(IR)_{max}}{K_M}[S] = \text{constant } [S] \qquad (3.13)$$

since substrate determinations are performed at constant enzyme concentration.

Experimentally, at least in most cases, the initial rate IR of substrate transformation is found directly proportional to the enzyme concentration. The same rate, however, follows "saturation kinetics" with respect to substrate concentration. This behavior is illustrated in Figure 3.1 and is taken into consideration in the generalized derivation given above and in equation 3.11.

### 3.3. THE MICHAELIS–MENTEN CONSTANT

Equation 3.9 shows that $K_M = [S]$ when the initial rate equals half the value of the maximum (saturation) rate. Since under initial rate conditions,

$$K_M = \frac{k_{-1} + k_2}{k_1}$$

if $k_{-1} \gg k_2$, $K_M = k_{-1}/k_1$ and the Michaelis–Menten constant becomes the dissociation constant for the enzyme–substrate complex ES. On the other hand, if $k_2 \gg k_{-1}$, $K_M = k_2/k_1$ and it should be considered as a kinetic constant.

It must be realized that IR is linearly related to $[S]$ when $K_M > [S]$ and as such a small $K_M$ value (indicating that the enzyme requires only a small amount of substrate to become saturated) connotes a short concentration range in calibration graphs prepared from initial rate measurements. The larger the value of $K_M$, the wider the useful concentration range for substrate determination.

Although not true for all enzymes, the substrate in front of which the enzyme has the lowest $K_M$ value is assumed to be the enzyme's natural substrate and should be preferred for enzyme concentration (activity) determinations that are performed under conditions of saturation kinetics. Under these conditions

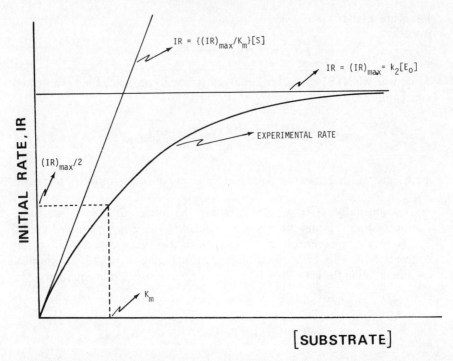

**Figure 3.1.** Reaction rate plotted against substrate concentration for an enzyme-catalyzed reaction obeying saturation kinetics. The saturation point, characterized by $(IR)_{max}$, can be interpreted as representing the situation in which all active sites are occupied by substrate at every time $t$ and substrate concentrations are larger than the minimum needed for realization of $(IR)_{max}$. From Ref. 11, reproduced with permission.

the initial rate of product formation follows zero-order kinetics (rate independent of substrate concentration) and is constant (linear) with time (Figure 3.2) as long as the enzyme concentration (activity) is the rate-limiting factor. Ideally, enzyme activity should be measured under these conditions.

### 3.4.  ENZYME ACTIVITY AND ENZYME CONCENTRATION

The catalytic activity of an enzyme is proportional to the number of operating active sites at the time of its measurement and is expressed in enzyme *activity units.* Although related to the analytical concentration of proteinaceous material, the catalytic activity may be different for different preparations of the same enzyme, so the analytical concentration is useless in describing enzyme activity. The situation is even worse than this: at different rates the activity of a

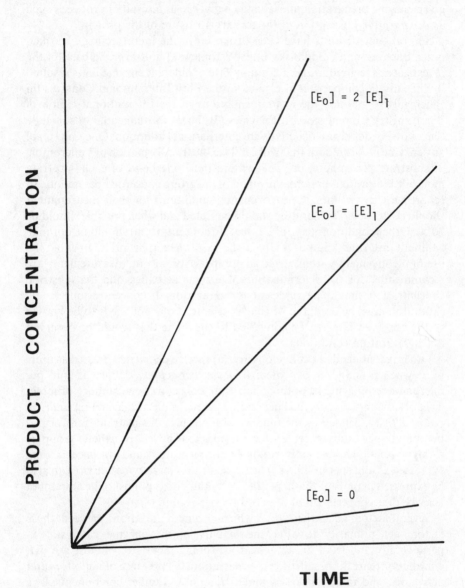

**Figure 3.2.** "Zero-order" product formation (where rate is independent of substrate concentration). $[E_0]$ = initial enzyme concentration (activity). This represents ideal conditions for enzyme activity determination.

given enzyme preparation may change with time as a result of processes such as *denaturation* (disruption of the tertiary structure of the protein).

Several enzyme units have been proposed in the literature (e.g., Karmen units, Buchner units, Bodansky units, Wroblewski units) on the basis of the investigator's technique and definition. Use of different enzyme units resulted in difficulty in comparison of results, so the Fifth International Congress on Biochemistry adopted the recommendations of the International Union of Biochemistry Commission on Enzymes (7). These recommendations include the following definition of activity for international adoption: One unit (U) of any enzyme is the amount that will catalyze the transformation of 1 micromole of substrate per minute, or, where more than one bond of each substrate molecule is attacked, 1 microequivalent of the group concerned per minute. In reporting activity units, the experimental conditions for their measurement should be defined. Temperature should be stated and when possible should be 30 °C. Other conditions (e.g., pH, substrate concentration) should be optimal. Milliunit (mU) and kilounit (kU) may be used when appropriate to avoid the use of inconveniently small or large numbers. Initial rate measurements are recommended for the determination of enzyme activities, and the substrate concentration should be sufficient for saturation. If concentrations below saturation must be used, the Michaelis–Menten constant needs to be known so the observed rate can be converted to the value that would be observed under saturation conditions.

Other recommended expressions are: (a) *specific activity*, expressed as units of enzyme per milligram of protein: (b) *molecular activity*, defined as units per micromole of enzyme at optimal substrate concentration (number of molecules of substrate transformed per minute per molecule of enzyme); (c) *catalytic center activity*, defined as the number of molecules of substrate transformed per minute per catalytic center (for enzymes that have a prosthetic group or catalytic center whose concentration can be measured); and (d) *concentration*, expressed as units per milliliter. The concentration of enzyme activity can also be expressed in units of "katal per liter"; "katal" corresponds to the amount of activity that converts 1 mol of substrate per second (8).

Several factors affect the rate at which enzymatic reactions proceed; these factors act primarily to alter the activity of the enzyme catalyst. Experimentally and from an analytical viewpoint, factors of interest are pH, enzyme concentration, substrate concentration, the presence of inhibitors and activators, and in certain cases ionic strength and buffer components. The effect and analytical applications of inhibition and activation are given in a separate chapter. The effect of enzyme and substrate concentration has been already discussed in the foregoing kinetic treatment of enzyme-catalyzed reactions in this chapter. The effect of temperature on enzyme-catalyzed reactions is about the same as for any other type of chemical reactions, and rules of thumb can be used such as a 10°C increase in temperature increases the rate by roughly 50 to 100%. Many enzymes, however, are affected (denatured) by only slightly elevated temperature. Most enzymes of animal origin, for instance, are denatured at temperatures above 40°C. Over a period of time, enzymes will be deactivated at even moderate (room) temperature, so it is a common practice to store enzymes and enzyme preparations at temperatures close to 5°C. Some enzymes, however, lose their activity if frozen.

As is the case for most reactions in aqueous solution, the rate of enzyme-catalyzed reactions is affected by pH changes. As with temperature (see Figure 3.3A), the rate profile as a function of pH takes a bell shape with a optimum pH (Figure 3.3B). Whenever possible, this optimum pH should be adopted for analytical applications of enzyme-catalyzed reactions. Consequently, hydrogen ion concentration also affects enzyme stability; this is of relevance in the storage of enzyme preparations. The optimum pH for different enzymes varies considerably. For instance, lipase (pancreas) exhibits an optimum pH of 8.0, while lipase (castor oil) has optimum pH between 4.0 and 5.0; the optimum pH for pepsin is 1.5 to 1.6 and that of trypsin is 7.8 to 8.7.

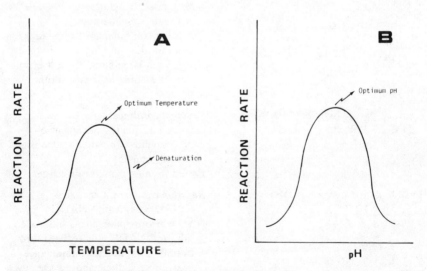

**Figure 3.3.** Typical effect of (A) temperature and (B) pH on the rate of an enzyme-catalyzed reaction.

### 3.5.   ANALYTICAL APPLICATIONS OF ENZYME-CATALYZED REACTIONS

#### 3.5.1.   Clinical Determinations of Enzyme Activities

Although hundreds of different enzymes have been identified in human tissue and about 50 in human serum (9), only 20 or so have been found, that relate to the diagnosis of disease. Table 3.2 gives an overview of some of them. The EC

**Table  3.2.  Typical Determinations of Enzymes Relevant to Clinical Diagnosis**

| Enzyme | Comments |
| --- | --- |
| Glutamic-oxaloacetic transaminase (GOT)[EC 2.6.1.1] | *Reaction catalyzed*: L-aspartate + 2-oxoglutarate $\rightleftharpoons$ oxaloacetate + L-glutamate<br>One of transaminases having greatest concentration in human body tissue; of importance in the metabolism of amino acids. Large amounts found in heart, liver, skeletal muscle, and kidney tissue; lesser amounts in brain, pancreas, and lung tissue. |
| Glutamic-pyruvic transaminase (GPT) [EC 2.6.1.2] | *Reaction catalyzed*: L-alanine + 2-oxoglutarate $\rightleftharpoons$ pyruvate + L-glutamate<br>The other transaminase having high concentration in human body tissue. Found in large amounts in liver but only in smaller amounts in other tissues. |
| Creatine phosphokinase (CPK) [EC 2.7.3.2] | *Reaction catalyzed*: creatine phosphate + adenosine 5′-diphosphate $\rightleftharpoons$ creatine + adenosine 5′-triphosphate<br>Found in muscle and brain tissue. |
| Lactate dehydrogenase (LDH) [EC 1.1.1.27] | *Reaction catalyzed*: L-lactate + NAD $\rightleftharpoons$ pyruvate + NADH<br>NAD = nicotinamide adenine dinucleotide, and NADH = its reduced form. Found in considerable abundance in heart, liver, kidney, and muscle tissue. |

From Ref. 11, reproduced with permission.

numbers following each enzyme cited in Tables 3.2 and 3.3 follow the classification suggested by the International Union of Biochemistry (IUB), which assigns a unique number to each enzyme. The interested reader can consult IUB publications for the meaning of each number (8, 10).

Other enzymes of interest in clinical diagnosis are some hydrolases (e.g., amylase, lipase, alkaline phosphatase, acid phosphatase leucine amino-peptidase), aldolase (a lyase that catalyzes the reversible decomposition of fructose diphosphate into dihydroxyacetone phosphate and D-glyceraldehyde phosphate), and phosphohexose isomerase (catalyzes the conversion of glucose-6-phosphate to fructose-6-phosphate).

### 3.5.2. Determination of Substrates by Means of Enzyme-Catalyzed Reactions (11)

In this type of application, the enzyme acts as an analytical reagent. These determinations are of relevance not only to clinical chemistry but also in many cases of industrial (pharmaceutical) and environmental problems. Table 3.3 provides some examples illustrating these types of applications.

A wealth of information pertinent to enzymes as analytical reagents can be found in the literature provided by commercial suppliers of enzymes. Reference 12, for instance, provides a comprehensive list giving reaction being catalyzed, characteristics of the enzyme (i.e., molecular weight, chemical composition, molar absorptivity, activators, inhibitors, selectivity, stability, optimum pH), practical directions for the determination of the enzyme, and pertinent references for slightly over 100 enzymes. Also included are properties of several related biochemicals of use in determinations involving enzymes.

### 3.6. CHEMICAL APPROACHES TO THE MONITORING OF ENZYME-CATALYZED REACTIONS

The chemistry of the enzyme-catalyzed reaction is what primarily dictates the approach, including instrumentation, to be used in monitoring the reaction rate. However, since many such reactions have some chemical characteristics in common, a few used ways of monitoring the reaction are worth mentioning.

Sometimes one of the reaction products shows an absorption spectrum different from the spectra of the substrate and other reactants or products involved in the reaction. This permits spectrophotometric monitoring of the course of the reaction in a simple manner. Nicotinamide adenine dinucleotide (NAD), a coenzyme for many dehydrogenase-catalyzed reactions, does not absorb photons in the 340-nm region, whereas its reduced form (NADH) does (Figure 3.4). This difference is widely used for following the progress of

**Table 3.3 Typical Determinations of Substrates Using Enzyme-Catalyzed reactions**

| Substrate | Comments |
|---|---|
| | *I. Clinical Applications* |
| Galactose | *Enzyme used*: galactose oxidase (EC 1.1.3.9)<br>*Reaction used*:<br>D-galactose + $O_2$ $\rightleftharpoons$ D-galacto-hexodialdose + $H_2O_2$<br>*Use*: diagnosis of galactosemia. |
| Glucose | *Enzyme used*: glucose oxidase (EC 1.1.3.4)<br>*Reaction used*:<br>glucose + $O_2$ + $H_2O$ $\rightleftharpoons$ gluconic acid + $H_2O_2$<br>*Use*: diagnosis of diabetes |
| Blood alcohol | *Enzyme used*: alcohol dehydrogenase (EC 1.1.1.1)<br>*Reaction used*:<br>Ethanol + NAD $\rightleftharpoons$ acetaldehyde + NADH<br>*Use*: law enforcement laboratories, treatment of alcoholism. |
| Uric acid | *Enzyme used*: uricase (EC 1.7.3.3)<br>*Reaction used*:<br>uric acid + $O_2$ + $H_2O$ $\rightleftharpoons$ allantoin + $CO_2$ + $H_2O_2$<br>*Use*: diagnosis of gout, chronic leukemia, lymphoma,<br>polycythemia, and myeloid metaplasia. |
| | *II. Industrial Applications* |
| Penicillins | *Enzyme used*: penicillinase (EC 3.5.2.6)<br>*Reaction used*:<br>penicillins + $H_2O$ $\rightleftharpoons$ penicilloic acid<br>*Use*: determination in pharmaceutical preparations and<br>fermentation broths. |
| Ethanol | *Enzyme used*: yeast alcohol dehydrogenase (EC 1.1.1.1)<br>*Reaction used*:<br>ethanol + NAD $\rightleftharpoons$ acetaldehyde + NADH<br>*Use*: determination in alcoholic beverages. |
| Maltose | *Enzyme used*: maltase ($\alpha$-glucosidase, EC 3.2.1.20)<br>*Reaction used*:<br>maltose + $H_2O$ $\rightleftharpoons$ 2D-glucose<br>Glucose determined by use of glucose oxidase.<br>*Use*: determination in food products (e.g., corn and malt syrups). |
| Sucrose | *Enzyme used*: invertase ($\beta$-fructosidase, EC 3.2.1.26)<br>*Reaction used*:<br>sucrose + $H_2O$ $\rightleftharpoons$ D-glucose + fructose<br>Glucose determined by use of glucose oxidase.<br>*Use*: determination in food products (e.g., fruit juices, ice cream,<br>condensed milk, frozen vegetables, syrups, fresh vegetables). |

**Table 3.3** (*continued*)

| Substrate | Comments |
|-----------|----------|
| Lactose | *Enzyme used*: lactase ($\beta$-galactosidase, EC 3.2.1.23)<br>*Reaction used*:<br>lactose + $H_2O \rightleftharpoons$ D-glucose + galactose<br>Glucose determined using glucose oxidase.<br>*Use*: determination in food products (e.g., milk, ice cream, yoghurt, buttermilk). |
| Catechol | *Enzyme used*: catechol 1,2-oxygenase (EC 1.13.1.1)<br>*Reaction used*:<br>catechol + $O_2 \rightleftharpoons$ dicarboxylic acid<br>Dicarboxylic acid results from cleavage of the ring between the two phenolic groups. $O_2$ levels monitored.<br>*Use*: determination in dyes, drugs, rubbers, antioxidants for lubricating oils, chemicals used in photography, cigarette smoke. |

### III. Environmental Application

| | |
|-----------|----------|
| Phenol | *Enzyme used*: tyrosinase (polyphenol oxidase, EC 1.14.18.1)<br>*Reaction used*:<br>Phenol + $O_2 \rightleftharpoons$ *o*-benzoquinone<br>$O_2$ levels monitored.<br>*Use*: determination in water and wastewater samples. |

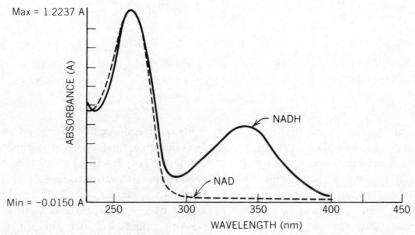

**Figure 3.4.** Difference in absorbance at 340 nm between nicotinamide adenine dinucleotide (NAD) and its reduced form (NADH). [NAD] = $1.25 \times 10^{-4}$ *M*; [NADH] = $2.50 \times 10^{-4}$ *M*.

reactions involving dehydrogenases by direct monitoring at 340 nm. A typical example is the determination of lactic dehydrogenase (LDH) (or lactic acid) by means of the following reaction:

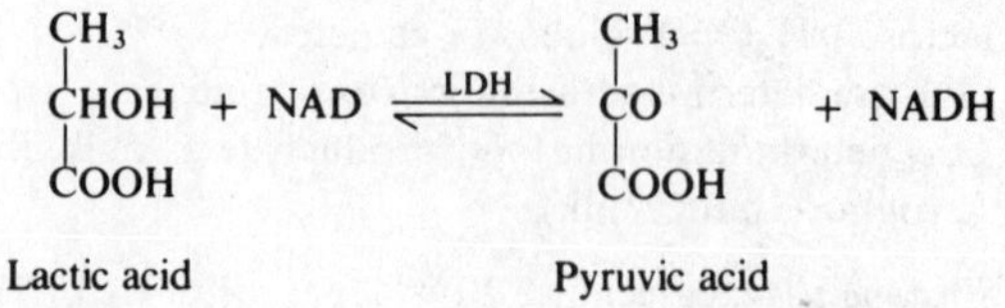

$$
\underset{\text{Lactic acid}}{\underset{\text{COOH}}{\overset{\text{CH}_3}{\text{CHOH}}}} + \text{NAD} \underset{\text{LDH}}{\rightleftharpoons} \underset{\text{Pyruvic acid}}{\underset{\text{COOH}}{\overset{\text{CH}_3}{\text{CO}}}} + \text{NADH}
$$

The structure of NAD is

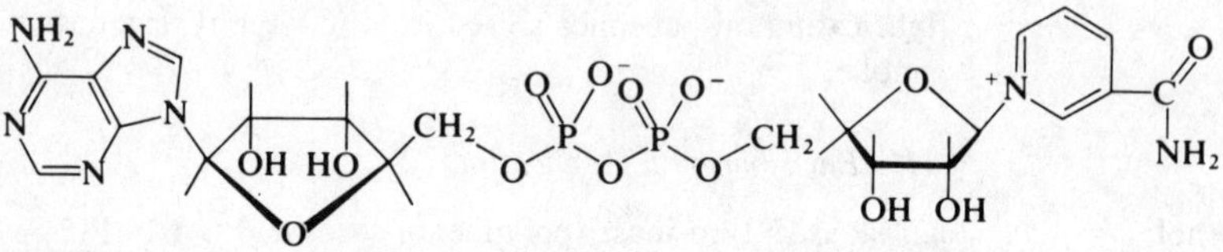

and it acts by accepting hydrogen (reversibly) at the 4 position of the nicotinamide ring:

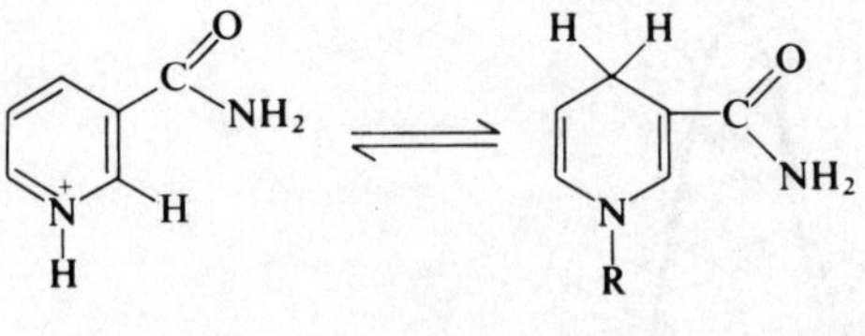

An increase in sensitivity and lower limits of detection can be achieved by coupling to an indicator reaction that affects the appearance or disappearance of a fluorescent species. Guilbault and Kramer (13, 14) described a simple and rapid method for the determination of the activity of dehydrogenases based on the conversion of nonfluorescent resazurin to resorufin, which is highly fluorescent. As little as $10^{-4}$ U/mL of lactic dehydrogenase, malic dehydrogenase, glutamate dehydrogenase, glucose-6-phosphate dehydrogenase, L-α-glycerophosphate dehydrogenase, and glycerol dehydrogenase have been determined, and as little as $10^{-9}$ $M$ resorufin can be detected:

$$\text{substrate} + \text{NAD} \xrightarrow{\text{dehydrogenase}} \text{oxidized substrate} + \text{NADH}$$

NADH +

Resazurin
(nonfluorescent)

Resorufin
($\lambda_{ex} = 540$ nm,
$\lambda_{em} = 580$ nm)

+ NAD

In other cases NAD is not involved in the reaction of interest but is used in a coupled reaction to facilitate monitoring by following the absorbance at 340 nm. An example of this is the determination of serum glutamic-oxaloacetic transaminase (GOT) by

$$\alpha\text{-ketoglutarate} + \text{aspartate} \underset{}{\overset{\text{GOT}}{\rightleftharpoons}} \text{glutamate} + \text{oxaloacetate}$$

$$\text{oxaloacetate} + \text{NADH} \underset{}{\overset{\text{MDH}}{\rightleftharpoons}} \text{malate} + \text{NAD}$$

where MDH = malic dehydrogenase.

The rate of decrease in the concentration of NADH is directly proportional to the GOT activity.

Such coupling of the main enzyme-catalyzed reaction with a second (indicator) reaction is a common practice in enzymatic methods. Perhaps the most popular of these couplings is to use an organic dye that reacts with a product (or the substrate) and then monitor the appearance or disappearance (or change) of color spectrophotometrically. Since reactions involving oxidases yield $H_2O_2$ as one of their products, the oxidizing power of $H_2O_2$ is exploited in the coupled scheme. This is aided by the action of a second enzyme (peroxidase) acting on the indicator reaction. A typical example is

$$\text{glucose} + O_2 + H_2O \underset{}{\overset{\text{glucose oxidase}}{\rightleftharpoons}} \text{gluconic acid} + H_2O_2$$

$$H_2O_2 + \text{dye (reduced form)} \underset{}{\overset{\text{peroxidase}}{\rightleftharpoons}} \text{dye (oxidized form)} + H_2O$$

A typical example of chemical species used in the coupled reaction is
$o$-dianisïdine (3,3'-dimethoxybenzidine), which is colorless in solution but in
oxidized form absorbs photons in the 460-nm region of the spectrum. Other
species used are $o$-tolidine and leuco bases of triphenylmethane dyes (e.g.,
leuco crystal violet).

The oxygen consumption and $H_2O_2$ formation in oxidase-catalyzed
reactions make these reactions amenable to amperometric detection of either
oxygen consumption or $H_2O_2$ formation. The use of amperometric detection
of oxidase-mediated reactions, with emphasis on applications using continuous-flow sample processing, has been reviewed by Gulberg et al. (15) and
Mottola et al. (16). Species such as glucose, D- and L-amino acids, uric acid,
amylase, maltose, sucrose, lactose, NADH, and serum lactate dehydrogenase
have been determined by amperometric monitoring of changes in dissolved
oxygen levels due to oxidase-catalyzed reactions.

Several commonly used enzyme-catalyzed reactions involve proton exchange with the background electrolyte. For these reactions pH sensing is the
simplest monitoring approach since it does not require additional coupling to
an indicator reaction. However, hydrogen ion sensing has some inconvenien-
ces since the pH change, if large enough, may inhibit enzyme activity. Buffering
species present in the medium may also affect the measurement. Keeping the
pH change small, however, allows reliable monitoring of the reaction rate.
Alternatively, a pH-stat approach may be adopted to circumvent the problem.
This approach consists of measuring the rate of acid (or base) added to a
potentiometrically equipped system in order to maintain the indicator
electrode at a constant potential (constant pH). An example of its use is the
decomposition of acetylcholine in presence of acetylcholinesterase, which
results in the formation of acetic acid. Acetylcholinesterase activity is very
sensitive to pH, and the medium must be held very close to pH 7.4 for the
reaction to proceed conveniently (17).

In certain cases, a careful design of buffer system and buffer capacity allows
a direct monitoring of pH (18–20). Enzyme-catalyzed reactions leading to the
release of ammonia, which in aqueous solution generates ammonium ion, are
amenable to direct monitoring by ammonium-ion-selective electrode probes.
Guilbault et al. (21), for instance, have described the enzymatic determination
of a series of substrates (urea, glutamine, asparagine, and glutamic acid) by
monitoring the reaction with a cation-selective electrode responding to
ammonium ion.

An interesting but rarely implemented exploitation of enzyme catalysis is an
amplification approach centered on the cycling of two reversibly interrelated
chemical species (e.g., the oxidized and reduced forms of a given chemical
structure) sequentially acting on two enzyme-catalyzed reactions. The cycling
results in an accumulation of product; if side reactions do not interfere with the

cycling process, the resulting amplification is significantly large. Cofactors such as NAD–NADH or NADP–NADPH are ideally suited to this approach. (NADP = nicotinamide adenine dinucleotide phosphate and NADPH = its reduced form.) An example, illustrated in Figure 3.5, is the cycling of NADP–NADPH between two enzyme-catalyzed reactions resulting in the accumulation of 6-phosphogluconate, which is determined by a third enzyme-catalyzed reaction (22). The sample, containing NADP, is added to a mixture containing three chemical species in nonlimiting concentrations—α-keto-glutarate, ammonium ion, and glucose-6-phosphate—as well as controlled amounts of the enzymes glutamate dehydrogenase (GDH) and glucose-6-phosphate dehydrogenase (G6PDH). Both 6-phosphogluconate and glutam-ate build up at a rate related to the original concentration of NADP in the sample. Once cycling has resulted in a measurable concentration of 6-phosphogluconate, the reaction series is quenched by heating the mixture, which denatures the protein and thus destroys the enzyme activity. The accumulated 6-phosphogluconate is then determined by addition of a known level of NADP and 6-phosphogluconate dehydrogenase (6PGDH):

$$6\text{-phosphogluconate} + NADP \xrightarrow{6PGDH} NADPH + \text{ribulose-5P} + CO_2 + H^+$$

If the enzyme concentrations were high enough, cycling rates of 20,000 per hour could be accomplished. If the NADPH formed in a first step is entered in a second cycling stage, another 20,000 cycles per hour could be implemented;

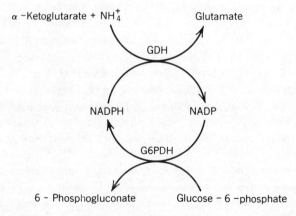

**Figure 3.5.** Amplification of NADP determination by means of enzymatic cycling (for explan-ation see text).

this would result in amplification factors of $10^8$, allowing detection of as little as $10^{-19}$ mol NADP. A short but thought-provoking account of the potentials of amplification by cycling enzyme-catalyzed reactions has been given by Lowry (23). The topic is also treated in some detail in a chapter of a book presenting a systematic scheme for the analysis of biological materials (24). In actuality, catalysis in itself is an amplification approach since substoichiometric amounts of catalyst yield a product accumulation by action of the catalytic cycle; this and other aspects of chemical amplification have been discussed by Blaedel and Boguslaski (25). An enzyme reactor (enzyme electrode) with a chemically amplified response for L-lactate has been proposed by Mizutani et al. (26). The electrode system consisted of an "oxygen electrode" (a platinum cathode and a gas-permeable membrane) and an immobilized enzyme layer in contact with the test solution and the gas-permeable membrane. The immobilized enzyme layer contained lactate oxidase (to oxidize L-lactate) and lactate dehydrogenase (to regenerate the L-lactate). The regeneration process permits oxygen consumption beyond the stoichiometric yield, resulting in an electrode response amplified 2 to 250 times depending on the characteristics of the immobilized enzyme layer (layer thickness, value of apparent $K_M$, etc.). A detection limit of $5 \times 10^{-9}$ $M$ is reported for L-lactate by this system.

## REFERENCES

1. T. P. Bennett and E. Frieden, *Modern Topics in Biochemistry*, Macmillan, London, 1969, p. 43.

2. H. Stetter, *Enzymatische Analyse*, Verlag-Chemie, Weinheim, N.Y., 1951.

3. G. G. Guilbault, *Enzymatic Methods of Analysis*, Pergamon Press, Oxford, 1970.

4. B. Harrow and A. Mazur, *Textbook of Biochemistry*, Saunders, Philadelphia, 1958, p. 109.

5. L. Michaelis and M. L. Menten, *Biochem. Z.*, **49**, 333 (1913).

6. G. E. Briggs and J. B. S. Haldane, *Biochem. J.*, **19**, 338 (1925).

7. *Report of the Commission on Enzymes*, International Union of Biochemistry Symposium Series, Pergamon Press, Oxford, 1961, Vol. 20, Chapter 9, p. 45.

8. "Enzyme Nomenclature," *Recommendations* (1972) *of the IUPAC and the International Union of Biochemistry*, Elsevier, Amsterdam, 1973, Chapter 4.

9. H. J. Zimmerman and J. B. Henry, in *"Clinical Diagnosis by Laboratory Methods"*, I. Davidsohn and J. B. Henry, Eds., Saunders, Philadelphia, 1969, p. 710.

10. *Report of the Commission on Enzymes*, International Union of Biochemistry Symposium Series, Pergamon Press, Oxford, 1961, Vol. 20, Appendix D, p. 61.

11. H. A. Mottola, *Analyst (London)*, **112**, 719 (1987).

12. *Worthington Enzyme Manual*, Lillian A. Decker, Ed., Worthington Biochemical Corporation, Freehold, N. J., 1977.

13. G. G. Guilbault and D. N. Kramer, *Anal. Chem.*, **36**, 2497 (1964).

14. G. G. Guilbault and D. N. Kramer, *Anal. Chem.*, **37**, 1219 (1965).

15. E. L. Gulberg, A. S. Attiyat, and G. D. Christian, *J. Autom. Chem.*, **2**, 189 (1980).

16. H. A. Mottola, Ch.-M. Wolff, A. Iob, and R. Gnanasekaran, "Potentiometric and Amperometric Detection in Flow Injection Enzymatic Determinations," in *Modern Trends in Analytical Chemistry*," Vol. 18, E. Pungor and I. Buzas, Eds., Akademiai Kiado, Budapest, 1984.

17. J. Jensen-Holm, H. H. Lausen, K. Milthers, and K. O. Moller, *Acta Pharmacol. Toxicol.*, **15**, 384 (1959).

18. J. Ruzicka, E. H. Hansen, A. K. Ghose, and H. A. Mottola, *Anal. Chem.*, **51**, 199 (1979).

19. K. D. Begum and H. A. Mottola, *Anal. Biochem.*, **142**, 1 (1984).

20. R. Gnanasekaran and H. A. Mottola, *Anal. Chem.*, **57**, 1005 (1985).

21. G. G. Guilbault, R. Smith, and J. Montalvo, *Anal. Chem.*, **41**, 600 (1969).

22. O. H. Lowry, J. V. Passonneau, D. W. Schulz, and M. K. Rock, *J. Biol. Chem.*, **236**, 2746, 2756 (1961).

23. O. H. Lowry, *Acc. Chem. Res.*, **6**, 289 (1973).

24. O. H. Lowry and J. V. Passonneau, *A Flexible System of Enzymatic Analysis*, Academic Press, New York, 1972, Chapter 8.

25. W. J. Blaedel and R. C. Boguslaski, *Anal. Chem.*, **50**, 1026 (1978).

26. F. Mizutani, T. Yamanaka, Y. Tanabe, and K. Tsuda, *Anal. Chim. Acta*, **177**, 153 (1985).

# METHODS OF DETERMINATION BASED ON MODIFIED CATALYZED REACTION RATES IN SOLUTION

The major impact of developing methods based on the modification of catalyzed reaction rates is in the determination of organic species. An increased concern with health-related problems and the desire to examine or re-examine sources of environmental change generated renewed interest in the determination of organic species in solution at low concentration levels. Modification of catalyzed reaction rates by these species is a way to make available the low limits of detection and competitive sensitivities of catalytic methods for species that are normally noncatalytic in their chemistry. Rate modification for the determination of metal ion species has been mostly limited to enzyme-catalyzed reactions.

The effect of different species in modifying catalytic effects and their analytical use, particularly with regard to metal ion catalysis, has received attention in a few reviews (1–5). Modification of enzyme-catalyzed reaction rates and their analytical use have also been reviewed in some detail (6, 7).

Analytical applications of modified rates are mainly the result of two opposite effects: (a) *inhibition*, and (b) *activation*. Both processes have been applied to enzyme as well as nonenzyme catalysis for analytical purposes.

## 4.1. MODIFICATION OF METAL-ION-CATALYZED SYSTEMS

The potential use of rate modifications in systems with metal ions as catalysts was recognized in Yatsimirskii's monograph on kinetic methods (8), but no description of chemical applications is given there.

From an analytical viewpoint, it is possible to distinguish three general types of effects when metal complexation results in the modification of reaction rates (9): (a) inhibition, (b) true metal-complex catalysis, and (c) promotion. True metal-complex catalysis and promotion are cases of activation. These effects depend on defining a catalyst as a chemical species that remains unaltered at the end of each catalytic cycle. Any of these modifying effects can be used for determining the modifying species or in some cases for increasing further the sensitivity in the determination of the metal catalyst itself (10).

### 4.1.1. Applications of Inhibition

*Inhibitors* are chemical species that combine with the catalyst to form some sort of complex. The complex either exerts less catalytic action than the free metal ion (*partial inhibition*), or complexation renders the catalyst completely inactive (*total inhibition*). In either case the effect on the reaction rate will be proportional to the concentration of inhibitor and can be used for its determination. The general practice followed for such determination has been to monitor the decrease in reaction rate in systems containing a constant amount of catalyst by adding increasing amounts of inhibitor to obtain working (calibration) curves; either by initial rate measurements or the method of tangents is used (11, 12). This approach offers relatively good limits of detection (10) but limited dynamic range of concentrations amenable to determination. Examples of this are the determination of ethylenediamino-$N,N,N',N'$-tetraacetic acid (EDTA) at the $10^{-6}$ $M$ level (11) and of amino acids (glycine, DL-serine, DL-phenylalanine, DL-glutamic acid, and L-arginine), also at the $10^{-6}$ $M$ level (12). The determination of EDTA is based on the total inhibition of the catalytic action of manganese ions on the malachite green–periodate indicator reaction. The determination of amino acids involves partial inhibition by formation of a 1:1 complex between copper (the catalyst) and the corresponding amino acid; the complex formed has less catalytic activity than copper itself on the oxidation of pyrocatechol violet by hydrogen peroxide.

If limits of detection can be somewhat sacrificed, a considerable extension in the corresponding concentration range determinable can be obtained by using titrimetry to make use of *catalytic end-point indication*, an approach ideally suited to the determination of inhibitors under conditions of total inhibition. The approach involves two consecutive reactions: (a) the *titration reaction*, in which a *catalytic titrant* added to the sample reacts rapidly and stoichiometrically with the sought-after species (the inhibitor); and (b) the *indicator reaction*, which involves the monitored species and under the given conditions, can occur at a noticeable rate, only when an excess of catalyst (titrant) is present in the system:

$$\text{Titrant} + \text{Sample} \longrightarrow \text{Products}$$
$$\text{Indicator reaction} \quad\text{---}\quad [\text{excess catalytic titrant}] \longrightarrow \text{Products}$$

The end point of the titration is indicated by the sudden increase or decrease of the monitored species, and the amount of catalyst needed to reach the end point is directly proportional to the amount of inhibitor present in the sample. Adding the catalyst at a constant rate and recording the change of signal with time gives a titration curve with a "pseudoinduction period" (Figure 4.1) of length proportional to the amount of inhibitor present (Figure 4.2). The end point, after the titration reaction has been for all practical purposes completed, can be located simply by direct extrapolation of the linear segments of a recorded titration curve as shown in Figure 4.1, from plots of observed rate coefficients (14), or from plots derived from the kinetics of the indicator reaction (15).

In 1960 Erdey and Buzas (16) employed the chemiluminescence from $H_2O_2$ and luminol as an indicator in the titration of EDTA with copper(II) ions and performed what seems to be the first application of catalytic end-point indication. Two years later Yatsimirskii and Fedorova (14) reintroduced in a direct manner the concept of catalytic end-point indication, and since then the topic has received attention in a few reviews (17–20). The wide range of inhibitor concentration that can be determined by use of catalytic end-point indication, as well as the good accuracy and precision afforded by this approach, is illustrated by typical dynamic ranges of $10^{-7}$ to $10^{-3}$ $M$ and uncertainties of 1 to 2% (18). Masking and demasking of metal ions using complexing agents permit a variety of titrimetric situations allowing the

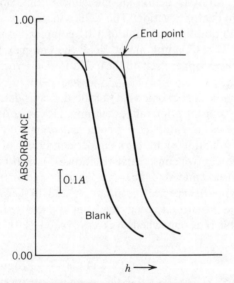

**Figure 4.1.** Typical shape of continuously recorded titration curves with catalytic end-point indication. Indicator reaction: malachite green $+ IO_4^-$; catalytic titrant: Mn(II). Titration of $1.05 \times 10^{-5}$ $M$ EDTA. [Reprinted with permission from H. A. Mottola, *Anal. Chem.*, **42**, 630, Copyright 1970 by the American Chemical Society.]

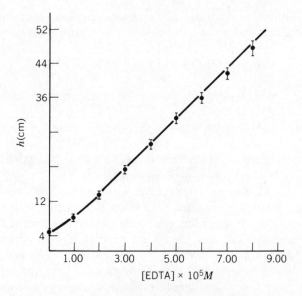

**Figure 4.2.** Typical working curve for EDTA determination by catalytic end-point indication; $h$ = length of pseudoinduction period. [Reprinted with permission from H. A. Mottola, *Anal. Chem.*, **42**, 630. Copyright 1970 by the American Chemical Society.]

determination of inhibitors by direct titration and, indirectly, of metal ions competing for the inhibitor (21–24). The back titration of noncatalytic metal ions requires that the formation constant of complexes of these ions with the inhibiting ligand be larger than the formation constants of the catalytic titrant and the inhibitor. A desirable requirement of the indicator reaction is a negligible uncatalyzed rate. If this requirement is not fulfilled, an alternative is to add the titrant mixed with one of the components of the indicator reaction to the other reactant (the monitored one) mixed with the inhibitor (the sought-for species). The concentration of the reactant mixed with the catalyst should be such that during the course of the titration this species is not rate-limiting (zero-order dependence on this reactant).

The course of the titration, and by extension the end-point indication, has been monitored by different procedures including visual indication (25), spectrophotometric indication (by far the most used approach) (13), potentiometric indication (26, 27), biamperometric indication (25), fluorometric indication (28), and thermometric indication (29). Thermometric indication has provided interesting applications. Acetic anhydride, for example, is commonly used to remove water from glacial acetic acid so that the latter can be used as a nonaqueous solvent in acid–base titrations. Vajgand and Gaál (29) have shown, however, that with glacial acetic acid containing about 2%

water and 8% acetic anhydride, tertiary amines can be conveniently titrated with thermometric end-point indication. The indicator reaction is

$$(CH_3CO)_2O + H_2O \longrightarrow 2CH_3COOH$$

An excess of perchloric acid present after the titration reaction.

$$\text{Organic Base} + HClO_4 \longrightarrow (\text{Protonated Base})^+ + ClO_4^-$$

catalyzes the indicator reaction. Thermometric sensing in catalytic end-point indication has been reviewed by Greenhow (30).

Computer simulation of titration curves with catalytic end-point indication has received some attention since 1973 when B. E. Simpson first simulated complexometric titrations of this type with spectrophotometric monitoring (31). More recently Gaál and Abramovic have authored a series of papers extending such studies to redox (32), precipitation (32), and acid–base reactions (33) with catalytic end-point indication. Their mathematical model takes into account the catalyst concentration before the equivalence point (34). Neglect of this results in an approximation valid for titration reactions exhibiting large formation constants (31). Studies of this kind, which consider the effect that different experimental conditions and physicochemical parameters may have on the shape of simulated titration curves, are helpful in selecting conditions for the development of new procedures and for optimizing those already proposed.

The shape of experimentally recorded titration curves may in some cases show activating and promoting effects. Under certain experimental conditions the complexes between manganese ions and certain aminopolycarboxylic acids [e.g., nitrilotriacetic acid, ethylene glycol bis(2-aminoethylether)tetraacetic acid, and EDTA] exert accelerating effects on the periodate ion oxidation of triphenylmethane dyes, as already mentioned in part. Such effects become clearly evident in the shape of the recorded titration curves (13).

### 4.1.2. Applications of Activation

Activation is a modification of the catalytic effect that, from an analytical point of view, yields a better sensitivity and lower limit of detection in a catalytic determination. In homogeneous catalysis an activator has been defined as a chemical species that does not catalyze the reaction but whose presence results in a considerable increase in the reaction rate (1). Activators have been classified in three groups, depending on the role of the catalyst, the catalyst–activator interaction, and the step of the reaction affected by the presence of the activator (1):

1. Activators affecting the catalyst-reactant interaction.
2. Activators participating in the regeneration of the catalyst.
3. Activators acting indirectly in the catalytic process.

True metal-complex catalysis is a typical example of *activation by affecting the catalyst–reactant interaction*. Metal ions catalyze the decarboxylation of oxaloacetic acid anion:

An increase in the effective charge of the metal ion should result in an increased catalytic activity. Bontchev and Michaylova (35) succeeded in increasing the effective charge of $Cu^{2+}$ by forming mixed ligand complexes with added organic solvents (e.g., ethanol, glycerol, butyric acid, and dioxane). Water of solvation was partially replaced by the organic solvent (activator), resulting in an increase of the effective charge of the metal ion catalyst. Complexing agents capable of forming catalyst–ligand $\pi$ bonds [e.g., pyridine, 1,10-phenanthroline, and tripyridyls (36)], also increase the effective charge in the metal ion catalyst. Activators can also affect the catalyst-substrate interaction by introducing steric factors (35), by providing an alternative mechanism for the catalyzed reaction (37) or bridging reagents in electron-transfer reactions (38), or by shifting the equilibrium in a reaction step (39).

Participation of activators in multielectron transfer reactions accounts for the second mode activation: *activators that accelerate the regeneration of the catalyst*. If, as happens with several transition metal ions, the metal ion catalyst can exist in several oxidation states, the activator can mediate the critical electron-transfer step and thus increase the overall rate of the catalyzed reaction (40). Theoretical concepts formulated to rationalize the kinetics of electron-transfer processes (41) have been successfully used to explain activation of this type (42) and can reveal potential activation. Facilitation of transition state formation is another mode of operation in this type of activation (43).

Chemical species that can participate in parallel processes affecting the catalytic rate represent *activators that act in an indirect way in the catalytic process*, the third mode of activation listed above. Examples are the effect of phenols on the catalytic oxidation of some arylamines (44) and the action of

hydroxycarboxylic and polycarboxylic acids in arylamine oxidations catalyzed by chromium(VI) (45).

As to the practical effect that activation may have on the sensitivity of
catalytic determinations, it can be mentioned, for example, that addition of
2,2'-bipyridine increases by 5000 times the sensitivity of silver determination
based on its catalysis of sulfanilic acid oxidation by persulfate (46) and that
manganese has been determined in the range of 4 to 40 ng in the presence of
nitrilotriacetic acid (NTA) in comparison to 0.1 to $2\mu g$ in its absence (47).

### 4.1.3.  Applications of Promoting Effects

In some instances the activation process is only transitory and is mainly
evident in earlier stages of the reaction. The transient nature of the rate-
modifying effect indicates that the substance acting as modifier (activator) is
rendered inactive or destroyed during the modified-catalytic reaction through
competitive reactions that halt the activating effect. As soon as the rate
modifier is destroyed (or inactivated), the overall rate tends toward that of the
reaction in its absence (Figure 4.3). This type of behavior seems better regarded
as *promotion* (9). As in true activation (e.g., true metal-complex catalysis), an
excess of promoter provides a system for the determination of catalyst with
improved sensitivity and limit of detection. The chromium(VI) oxidation of
tris(1,10-phenanthroline)iron(II) complex (ferroin) in sulfuric acid medium is
affected by a series of chemical species that accelerate the rate of oxidation in a
transient manner (48). In all cases, these effects are observed in the earlier
portions of the reaction profile. Conversion of the rate-modifying species to an
inactive form by decomposition, or more often by complexation with Cr(III)
produced in the main reaction, seems to account for the lack of a sustained
'catalytic cycle'. By correct choice of reaction conditions, microgram per
milliliter levels of oxalic acid, citric acid, vanadium(IV), arsenic(III),
chromium(VI), hexacyanoferrate(III), and milligrams per milliliter of
molybdenum(VI), can be determined (48). Titanium(III) was found to promote
the oxidation of iodide by hydrogen peroxide (49). The proposed promoting
mechanism involves a pre-equilibrium step between a monomeric and a
dimeric form of titanium(III), followed by free-radical formation in which the
monomeric form reacts with $H_2O_2$ to form the active radical $\cdot OH$, which
oxidizes the iodide ion. In the course of these processes the active titanium(III)
is converted to inactive titanium(IV). Stopped-flow measurements in tha
presence of 0.08 $M$ hydrochloric acid make it possible to determine
titanium(III) at the $10^{-5}$ $M$ level. The half-life of the promoting effect under the
specified conditions is estimated as 2 to 3 s.

Sodium thiobarbitone promotes the oxidation of pyrocatechol violet by
$H_2O_2$ catalyzed by copper(II) (50). Sodium thiobarbitone is a derivative of

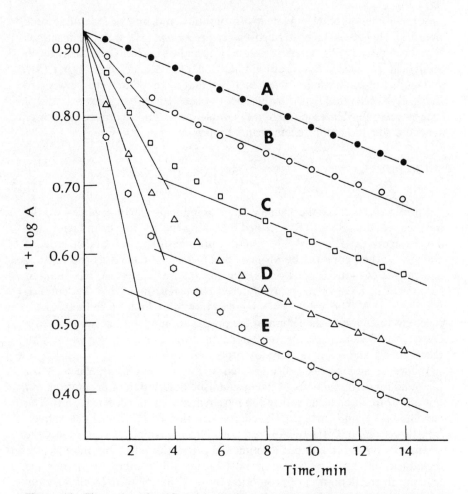

**Figure 4.3.** First-order plots for the reaction of tris(1,10-phenanthroline)iron(II) with chromium(VI), showing oxalic acid promotion. [Cr(VI)] = $8.33 \times 10^{-5}$ $M$; [ferroin] = $7.60 \times 10^{-5}$ $M$; [sulfuric acid] = 0.25 $M$. Concentration of oxalic acid added: (A) 0; (B) $1.00 \times 10^{-5}$ $M$; (C) 2.00 $\times 10^{-5}$ $M$; (D) $3.00 \times 10^{-5}$ $M$; (E) $4.00 \times 10^{-5}$ $M$. [Reprinted with permission from V. V. S. Eswara Dutt and H. A. Mottola, *Anal. Chem.*, **46**, 1090. Copyright 1974 by the American Chemical Society.]

thiobarbituric acid and is used as a short-action anesthetic. By using the method of tangents, sodium thiobarbitone has been determined at the $10^{-6}$ $M$ level. The effect of sodium thiobarbitone is seen only during the first few minutes of reaction; after about 4 min the overall rate becomes the same as the rate for the presence of copper(II) alone.

A phenomenon closely related to promotion and known for a long time appears in what has been termed *induced reactions* (51). Induced reactions have been discussed in detail by Skrabel (52) and Csanyi (53), and their role in analytical chemistry has been summarized by Laitinen and Harris (54). Induction appears when some reactions that take place very slowly are markedly accelerated by the simultaneous occurrence of another reaction of measurable rate. The reaction of measurable rate is said to induce the slower reaction. The induction scheme can be represented as follows:

$$R + I \rightarrow P \tag{4.1}$$
$$R + Ac \rightarrow P' \tag{4.2}$$

Reaction 4.1 is called the primary, main, or inducing reaction, while 4.2 is the induced reaction. Species R is termed the acting species, I is the inductor, and Ac the acceptor; P and P' are reaction products. The amount of induced change can be expressed by an *induction factor* $F_i$, defined as the ratio of equivalents of induced reaction to equivalents of primary reaction. Induced reactions can be devided in (a) "induced chain reactions" and (b) "coupled reactions" (54). The value of $F_i$ increases without limit in induced chain reactions (increase in the length of the propagation chain) but approaches a definite small value (e.g., 1, 2, or $\frac{1}{2}$) in coupled reactions. In most cases, both reactions 4.1 and 4.2 involve redox processes.

Induced reactions can be considered as due to promoting effects. Most likely, all induced reactions fit the mechanistic description of promotion, but not all promoting effects satisfy the requirements of induced reactions. The vanadium(IV) and arsenic(III) effects on the tris(1,10-phenanthroline)-iron(II)–chromium(VI) indicator reaction (48), for example, fit equally well the requirements of both induced reaction and promoting effect, but the effects of oxalic acid, citric acid, hexacyanoferrate(II), and molybdenum(VI) on the same reaction fit the definition of promotion but not that of an induced reaction.

### 4.1.4. Oscillating Reactions

A unique type of rate modification(s) occurs in the *oscillating reactions*, which are coupled autocatalytic systems in which repeated undamped oscillations occur in the concentrations of the reactants (55). Chemical oscillations are of theoretical interest and pertinent to some biological processes; however, their analytical application has been very limited. An extensively studied system that has received some analytical attention is the catalytic oxidation of citric, malonic, and succinic acids by bromate ion. Metal ions that can exchange a single electron during the course of the oscillations and have a standard redox potential of 0.9 to 1.6 V catalyze these oxidations. Mechanistically these processes are complex; for the bromate–malonic acid reaction catalyzed by Ce(IV)–Ce(III), a 10-step process for the mechanism has been proposed backed by detailed kinetic and thermodynamic arguments (56). Analytically important is the fact that the number of cycles per unit of time is proportional to initial reactant concentrations. However, this relationship is obeyed well for only a short time after the initiation of the reaction(s); when reagent concentrations decrease, so does the cycle frequency. Temperature control is very critical for successful analytical use of this relationship. An example of such use is the determination of ruthenium based on the increase of the period frequency in the oxidation of malonic acid catalyzed by cerium in presence of the Ru(III)–Ru(IV) redox couple (57). As little as $0.01 \, \mu g/mL$ Ru can be determined with excellent reproducibility, and small concentrations of Pt, Rh, and Ir do not interfere.

## 4.2. MODIFICATION OF ENZYME-CATALYZED SYSTEMS

The rates of enzyme-catalyzed reactions can be increased by the presence of *activators* or considerably decreased by the presence of *inhibitors*. Some metal ions, anions of some organic acids, and proteins can act as activators. Inhibition is caused by formation of either an enzyme–inhibitor complex or an enzyme–substrate–inhibitor complex. It is of interest to note that cofactors (see Chapter 3) are activators that make the enzyme-catalyzed reaction possible.

### 4.2.1. Activation

Activation of enzyme-catalyzed reaction exhibits a high degree of selectivity because the mediation of the activator between enzyme and substrate involves binding sites with specifically tailored sizes and shapes or electric fields with a unique intensity. Calcium ion, for instance, interacts with certain negatively charged sites in enzymes such as trypsin. This interaction has no direct effect on the activity of trypsin, but it helps to maintain the native configuration and protects against denaturation. Certain activators, called *allosteric*, interact at special sites away from the active one; the new pattern of forces resulting from this interaction forces the enzyme to assume its correct configuration for catalysis. The overall result of these types of interaction is an enhanced activity in the presence of these activators.

The initial rate dependence in the presence of activator parallels the one discussed in Chapter 3 for the dependence on substrate concentration. At low concentrations of the activator the initial rate is directly proportional to the concentration of the activator; at relatively high values of the same activator the rate is independent of such concentration. This is not surprising since conceptually some substrates may act as activators.

Few analytical applications of enzyme activation can be found in the literature. Baum and Czok (58) developed a method for the determination of magnesium in plasma based on its activation of isocitric dehydrogenase. As an example of the high degree of selectivity exhibited by enzyme activation, Blaedel and Hicks (59) have indicated that only magnesium and manganese(II) noticeably activate this enzyme.

A unique situation for analytical exploitation of enzyme activation is offered by some metalloenzymes. In such enzymes, metals act as (a) electrophilic catalysts stabilizing the negative charges formed, (b) creators of potent nucleophiles by metal-bound hydroxide ions, and (c) electron conduits in redox reactions (60). In other examples the situation is more complex, as in the action of cobalt in vitamin $B_{12}$ reactions; cobalt forms covalent bonds with the substrate (60).

The possibility of removing the metal (by means of a complexing agent) from the metalloenzyme to obtain the corresponding apoenzyme provides a very selective and sensitive reagent for the determination of the metal ion by enzyme reactivation (61). Exploiting this approach, Lehky and Stein (62) developed a method that under ideal conditions would allow determination of less than 1 pg of zinc per milliliter. The selectivity of their method is also attractive; 0.1 ng mL of Zn can be determined with good accuracy even in the presence of a 10- to 100-fold excess of most other metals. The reagent used by Lehky and Stein was obtained by removing zinc from pig kidney aminopeptidase with EDTA. A limit of detection for zinc of 10 ng mL$^{-1}$ was reported by Kobayashi et al. (63) by its restoration of the catalytic activity of apocarbonic anhydrase. Carbonic anhydrase is a metalloenzyme containing 1 mol of tightly bound zinc per single polypeptide chain with a molecular weight of about 29,000. The zinc-free apoenzyme was prepared from bovine carbonic anhydrase (EC 4.2.1.1*) dissolved in water, dialyzed with 0.01 $M$ 1,10-phenanthroline and sodium acetate buffer (pH = 5.0) for 5 to 10 days. After further dialysis with $10^{-4}$ $M$ EDTA and 0.06 $M$ Tris buffer (pH = 7.8) for half a day, the chelating agents were removed by dialysis with water. This procedure yielded a product in which the zinc was reduced to less than 3% of the original content in the metalloenzyme. Determinations were based on rate measurement during the first 5 min after mixing. This approach was used to determine zinc in fruit juices and water samples.

Donangelo and Chang (64) have similarly determined zinc in plasma and serum using the apoenzyme obtained by removal of the zinc from *Escherichia*

---

\* The EC numbers given here and throughout the monograph are from the International Union of Biochemistry (IUB) classification of enzymes.

*coli* alkaline phosphatase (EC 3.1.3.1) with nitrilotriacetic acid (NTA). The original enzyme is a dimeric protein containing zinc and magnesium (molecular weight = 80,000 to 89,000). Magnesium does not activate the enzyme but enhances activity of the enzyme containing two gram–atoms of zinc. In the method described, which takes slightly over 1 h per determination, the apoenzyme is incubated with a serum or plasma sample. Some of the enzyme is reactivated by the zinc in the sample, the extent of reactivation being proportional to the available zinc.

### 4.2.2. Inhibition

Besides being irreversibly inactivated by heat or chemical reagents, enzymes may be reversibly inhibited by the noncovalent binding of inhibitors. There are three main types of inhibition: (a) competitive, (b) noncompetitive, and (c) uncompetitive. Most theories about the mechanism of enzyme inhibition are based on the existence of the enzyme–substrate complex. The lock-and-key analogy introduced by Fischer (65) and based on the idea of complementarity in enzyme–substrate interactions can be used to illustrate the three main types of inhibition. This is shown in Figure 4.4.

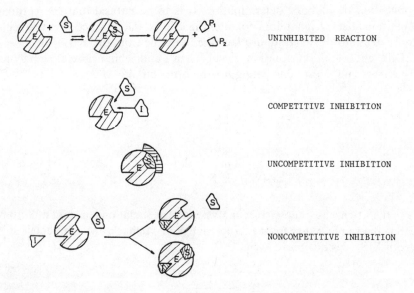

**Figure 4.4.** Graphical representation of the main types of enzyme inhibition using the lock-and-key analogy E, enzyme; S, Substrate; I, inhibitor; $P_1$, $P_2$, products. [from *Ion Selective Electrode Reviews*, Pergamon Press, 1985, Vol. 7, p. 50. Reproduced with permission.]

In *competitive inhibition*, the inhibitor I and the substrate S compete for the active site. Thus in the simple Michaelis–Menten mechanism an additional equilibrium must be considered:

$$S + E \underset{k_{-1}}{\overset{k_1}{\rightleftharpoons}} [ES] \xrightarrow{k_2} P + E$$

$$I \Big\Updownarrow$$

$$[EI]$$

We solve the equilibrium and rate equations by taking into consideration that $[E]_0 = [ES] + [EI] + [E]$:

$$IR = \frac{k_2[E]_0[S]}{[S] + K_M(1 + [I]/K_i)}$$

where IR = initial rate, $K_i$ = equilibrium constant for the inhibition reaction $E + I \rightleftharpoons EI$. The effect is an increase of the Michaelis–Menten constant by a factor of $1 + [I]/K_i$. For fixed [S], increasing concentrations of inhibitor proportionally decrease the initial rate and this is the basis for quantitative application of inhibitor determination. It is to be noticed that competitive inhibition affects $K_M$ but does not alter the value of $(IR)_{max}$ since infinitely high concentrations of substrate displace the inhibitor from the active site.

Different inhibition situations occur when I and S bind simultaneously to the enzyme instead of competing for the same binding site:

$$E + S \rightleftharpoons ES \xrightarrow{k_2} P + E$$

$$I \Big\Updownarrow \qquad I \Big\Updownarrow$$

$$EI + S \qquad EIS$$

In the Michaelis–Menten mechanism, if the dissociation constant of S from EIS is the same as that from ES (no change in Michaelis–Menten constant), and if EIS does not react, then

$$IR = \frac{k_2/(1 + [I]/K_i)}{[S] + K_M}$$

This results in *noncompetitive inhibition*, in which $K_M$ is unaffected but $k_2$ is lowered (by the presence of the inhibitor) by a factor of $1 + [I]/K_i$. More commonly, the dissociation constant for liberation of S from EIS is different from that of ES, and both $K_M$ and $k_2$ are affected by the presence of the inhibitor. This case of inhibition is sometimes called *mixed inhibition*.

When the inhibitor binds to the enzyme–substrate complex but not to the free enzyme, the inhibitory effect is called *uncompetitive*, and the steady-state approximation in the Michaelis–Menten mechanism leads to

$$IR = \frac{(IR)_{max}}{1 + ([I]/K_i)} \frac{[S]}{[S] + K_M(1 + [I]/K_i)}$$

The presence of the inhibitor lowers both $(IR)_{max}$ and $K_M$.

Working (calibration) curves for the determination of inhibitors are commonly prepared by plotting the decrease of the initial rate against the inhibitor concentration. Typically these plots follow an exponential function with an approximate linear range extending from 0 to about 65% inhibition.

A large number of common inorganic cations and a few anions (e.g., $Ag^+$, $Al^{3+}$, $Be^{2+}$, $Bi^{3+}$, $Ce^{3+}$, $Cd^{2+}$, $Co^{2+}$, $Cu^{2+}$, $Fe^{2+}$, $Hg^{2+}$, $In^{3+}$, $Mn^{2+}$, $Ni^{2+}$, $Pb^{2+}$, $Zn^{2+}$, $CN^-$, $Cr_2O_7^{2-}$, $F^-$, and $S^{2-}$), have been determined by methods exploiting enzymatic inhibition (66). However, determination of organic species used as pesticides is the most relevant application of enzymatic inhibition. Dichlorodiphenyltrichloroethane (DDT) has been determined by inhibition of carbonic anhydrase (67). Some organophosphorus compounds can be determined with high selectivity and sensitivity by their inhibition of cholinesterases (68). Lipase inhibition provides a sensitive method for heptachlor, aldrin, lindane, and 2,4-dichlorophenoxyacetic acid (2,4-D) (6). Aldrin, heptachlor, and methyl parathion can be determined by exploiting alkaline phosphatase inhibition (69).

## REFERENCES

1. P. R. Bontchev, *Talanta*, **19**, 675 (1972).

2. M. Kopanica and V. Stara, *Chem. Listy*, **67**, 952 (1973).

3. H. A. Mottola, *Anal. Chim. Acta*, **71**, 443 (1974).

4. D. P. Nikolelis and T. P. Hadjiioannou, *Rev. Anal. Chem.*, **4**, 81 (1979).

5. G. A. Milovanovic, *Microchem. J.*, **28**, 437 (1983).

6. G. G. Guilbault, in *MTP International Review of Science: Physical Chemistry*, Series One, Part 12, T. S. West, Ed., Butterworths–University Park Press, Oxford, 1973, Chapter 5.

7. A. Townshend, *Process Biochem.*, **8**, 22 (1973).

8. K. B. Yatsimirskii, *Kinetic Methods of Analysis*, Pergamon, Oxford, 1966, Chapter 5.

9. A. E. Martell, *Pure Appl. Chem.*, **17**, 129 (1968).

10. H. A. Mottola and C. R. Harrison, *Talanta*, **18**, 683 (1971).

11. H. A. Mottola and H. Freiser, *Anal. Chem.*, **39**, 1294 (1967).

12. T. J. Janjic and G. A. Milovanovic, *Anal. Chem.*, **45**, 390 (1973).

13. H. A. Mottola, *Anal. Chem.*, **42**, 630 (1970).

14. K. B. Yatsimirskii and T. I. Fedorova, *Dokl. Akad. Nauk SSSR*, **143**, 143 (1962); *Chem. Abstr.*, **57**, 2821c (1962).

15. H. A. Mottola and H. Freiser, *Anal. Chem.*, **40**, 1266 (1968).

16. L. Erdey and I. Buzas, *Anal. Chim. Acta*, **22**, 524 (1960).

17. H. A. Mottola, *Talanta*, **16**, 1267 (1969).

18. T. P. Hadjiioannou, *Rev. Anal. Chem*, **3**, 82 (1976).

19. H. Weisz and S. Pantel, *Anal. Chim. Acta*, **62**, 361 (1972); *Fresenius' Z. Anal. Chem.*, **264**, 389 (1973).

20. F. F. Gaál, *Analyst (London)*, **112**, 739 (1987).

21. H. Weisz and T. Kiss, *Fresenius' Z. Anal. Chem.*, **249**, 302 (1970).

22. D. Klockow and L. Garcia Beltran, *Fresenius' Z. Anal. Chem.*, **249**, 304 (1970).

23. T. F. A. Kiss, *Mikrochim. Acta*, 847 (1973).

24. M. Ternero, F. Pino, M. D. Perez Bendito, and M. Valcarcel, *Anal. Chim. Acta*, **109**, 401 (1979).

25. H. Weisz and S. Pantel, *Anal. Chim. Acta*, **62**, 361 (1972); **64**, 389 (1973).

26. M. M. Thimotheou-Potamia, M. A. Koupparis, and T. P. Hadjiioannou, *Mikrochim. Acta*, 433 (1982).

27. B. F. Abramovic and F. F. Gaál, *Microchem. J.*, **32**, 226 (1985).

28. A. Moreno, M. Silva, M. D. Perez Bendito, and M. Valcarcel, *Analyst (London)*, **109**, 249 (1984).

29. V. J. Vajgand and F. F. Gaál, *Talanta*, **14**, 345 (1967).

30. E. J. Greenhow, *Chem. Rev.*, **77**, 835 (1977).

31. B. E. Simpson, "Some studies of simulation and analysis of indicator reactions for catalytic determinations," M.Sc. thesis, Oklahoma State University, 1973.

32. B. F. Abramovic, F. F. Gaál, and D. Z. Paunic, *Talanta*, **32**, 549 (1985).

33. F. F. Gaál and B. F. Abramovic, *Talanta*, **32**, 559 (1985).

34. F. F. Gaál and B. F. Abramovic, *Talanta*, **31**, 987 (1984).

35. P. R. Bontchev and V. Michaylova, *J. Inorg. Nucl. Chem.*, **29**, 2945 (1967).

36. J. V. Rund and R. A. Plane, *J. Am. Chem. Soc.*, **86**, 367 (1964).

37. P. R. Bontchev and B. G. Jeliazkova, *Inorg. Chim. Acta*, **5**, 75 (1971).

38. H. Taube, *Can. J. Chem.* **37**, 129 (1959).

39. N. Uri, *J. Phys. Colloid Chem.*, **53**, 1070 (1949).

40. A. A. Alexiev and P. R. Bontchev, *Mikrochim. Acta*, 13 (1970).

41. R. A. Marcus, *J. Chem. Phys.*, **24**, 966 (1956).

42. P. R. Bontchev and A. A. Alexiev, *J. Inorg. Nucl. Chem.*, **32,** 2237 (1970).

43. P. R. Bontchev and Z. Mladenova, *Mikrochim. Acta*, 427 (1967).

44. P. R. Bontchev, *Mikrochim. Acta*, 577 (1962).

45. E. I. Yassinskene and E. B. Bilidene, *Zh. Anal. Khim.*, **23,** 143 (1968).

46. P. R. Bontchev, A. A. Alexiev, and B. Dimitrova, *Talanta*, **16,** 597 (1969).

47. D. P. Nikoleis and T. P. Hadjiioannou, *Anal. Chem.*, **50,** 205 (1978).

48. V. V. S. Eswara Dutt and H. A. Mottola, *Anal. Chem.*, **46,** 1090 (1974).

49. E. Piemont, J. L. Leibenguth, and J.-P. Schwing, *Bull. Soc. Chim. Fr.*, 1254 (1979).

50. G. A. Milovanovic and M. J. Protolipac, *Bull. Soc. Chim. Beograd.*, **46,** 685 (1981).

51. F. Kessler, *Poggendorf's Ann.*, **95,** 224 (1855); **119,** 218 (1863).

52. A. Skrabel, *Die Induzierten Reaktionen; ihre Gesichte und Theorie*, F. Enkke, Stuttgart, 1908.

53. L. J. Csanyi, in *Comprehensive Chemical Kinetics*, Vol. 7, C. H. Bamford and C. F. H. Tipper, Eds., Elsevier, Amsterdam, 1972, Chapter 5, pp. 510–580.

54. H. A. Laitinen and W. E. Harris, *Chemical Analysis*, 2nd ed., McGraw-Hill, New York, 1975, Chapter 15, pp. 297–303.

55. R. J. Field, *J. Chem. Educ.*, **49,** 308 (1972).

56. R. J. Field, E. Koros, and R. M. Noyes, *J. Am. Chem. Soc.*, **94,** 8649 (1972).

57. L. P. Tikhonova, L. N. Zakrevskaya, and K. B. Yatsimirskii, *Zh. Anal. Khim.*, **33,** 191 (1978).

58. P. Baum and R. Czok, *Biochem. Z.*, **332,** 121 (1959).

59. W. J. Blaedel and G. P. Hicks, in *Advances in Analytical Chemistry and Instrumentation*, Vol. 3, C. N. Reilley, Ed., Interscience, New York, 1964, p. 105.

60. A. Fersht, *Enzyme Structure and Mechanism*, Freeman, Belfast, 1977, pp. 49–51.

61. A. Townshend and A. Vaughan, *Anal. Chim. Acta*, **49,** 366 (1970).

62. P. Lehky and E. A. Stein, *Anal. Chim. Acta*, **70,** 85 (1974).

63. K. Kobayashi, K. Fujiwara, H. Haraguchi, and K. Fuwa, *Bull. Chem. Soc. Jpn.*, **52,** 1932 (1979).

64. C. M. Donangelo and G. W. Chang, *Clin. Chim. Acta*, **113,** 201 (1981).

65. E. Fischer, *Ber. Dtsch. Chem. Ges.*, **27,** 2985 (1894).

66. G. G. Guilbault, *Enzymatic Methods of Analysis*, Pergamon, Oxford, 1970.

67. H. Keller, *Naturwissenschaften*, **39,** 109 (1965).

68. D. N. Kramer and R. M. Gamson, *Anal. Chem.*, **29** (12), 21A (1957).

69. G. G. Guilbault, M. H. Sadar, and M. Zimmer, *Anal. Chim. Acta*, **44,** 361 (1969).

CHAPTER

5

# ANALYTICAL APPLICATIONS OF HETEROGENEOUS CATALYSIS

Electrocatalysis (coupled cyclic chemical reactions involved in electrode processes) and the use of immobilized enzymes constitute the two main areas of heterogeneous catalysis with analytical applications. However, a few other examples are worth mentioning. The in situ generation of a heterogeneous catalytic entity in a homogeneous system, for example, constitutes one such application (1). Palladium and sulfur-containing inhibitors have been determined by generating Pd(0) from Pd(II)-containing solutions by reaction with hypophosphite:

$$Pd(II) + H_2PO_2^- + H_2O \longrightarrow Pd(0) + H_2PO_3^- + 2H^+$$

The metallic palladium formed then acts as a catalyst in the following two reactions:

$$H_2PO_2^- + H_2O \xrightarrow{\text{Pd(0)}} H_2PO_3^- + H_2$$

$$H_2 \xrightarrow{\text{Pd(0)}} 2H^*$$

The activated hydrogen H* on the palladium surface is a powerful reducing agent and operates on an indicator reaction making use of an organic dye:

$$\text{dye (oxidized form)} + H^* \longrightarrow \text{dye (reduced form)}$$

An activation step seems to be responsible for the presence of an induction period whose measurement provides the basis for determination. Using the triphenylmethane dye erioglaucine and photometric monitoring permitted the determination of submicrogram amounts of palladium and fractions of micrograms-per-milliliter concentrations of inhibitors such as iodide, sulfide, cyanide, thiocyanate, thiosulfate, hexacyanoferrate(II), and several organophosphorothioate pesticides (e.g., diazinon, disulfoton, *O,O*-dimethyl *O*-[4-(methylthio)-*m*-tolyl]phosphorothioate, malathion, *O,O*-diethyl *O-p*-nitrophenyl phosphorothioate, and *O,O*-dimethyl *S*-[4-oxo-1,2,3-

benzothiazin-3(4H)-ylmethyl]phosphorothioate (1). It is of interest that the most common type of inhibition in homogeneous catalysis is due to complex formation, and stoichiometric or excess amounts of the inhibitor are needed for total inhibition. Substoichiometric amounts of inhibitors can produce total inhibition in heterogeneous catalysis generated in situ. Another application of this type of heterogeneous catalysis has been reported by Sánchez-Pedreño and co-workers (2). Gold(III) has an initial inhibitory effect on the reaction between toluidine blue and hypophosphite catalyzed by in situ generation of Pd(0) in the reaction shown above. This permits determination of gold at the microgram level since increasing concentrations of gold increase the length of the induction period in a proportional manner; that is, the length of the induction period is a linear function of the [Au(III)]/[Pd(II)] ratio, within certain limits.

## 5.1.  HETEROGENEOUS CATALYZED ELECTRODE REACTIONS

Electrode processes can be considered heterogeneous chemical reactions taking place at the interface of a metal and an electrolyte, accompanied by the transfer of electrons through this interface (3). In some electrode processes the current intensity is affected by the kinetics of chemical reactions that occur in the homogeneous (electrolyte) solution surrounding the electrode, and a catalytic effect (and cycle) becomes operative. These heterogeneous catalyzed electrode processes involve chemical reactions either prior to or after the electron transfer at the electrode (4). At high rates of charge transfer no appreciable concentration gradients are produced in the vicinity of the metal electrode. The current is then controlled (partially or totally) by diffusion of the electroactive species from the bulk of the solution to the electrode–solution interface. Such diffusion control of the current does not always prevail, however; sometimes the kinetics of the chemical reaction taking place in the vicinity of the metal electrode controls the process. Currents developed under these conditions are termed *kinetic currents*, and the homogeneous chemical reaction is considered to be coupled to the electrode process.

### 5.1.1.  Coupled Chemical Reactions

The various types of kinetic currents have been discussed in detail by Guidelli (5); Figure 5.1 illustrates the classification and graphical representation provided by Guidelli. These schemes include electrode processes coupled with preceding, parallel, and subsequent reactions. Coupled chemical reactions are classified according to the overall chemical process and the sequence of steps, and letters are used to signify the nature of the process (6). For example, E

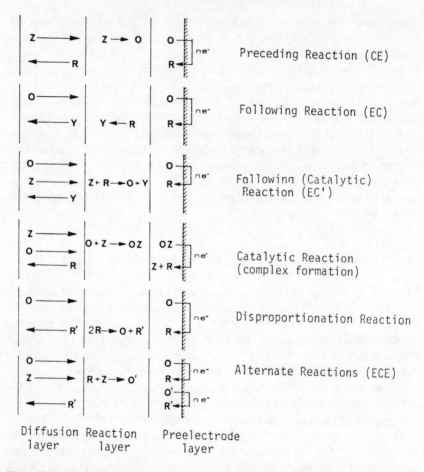

**Figure 5.1.** Chemical reactions coupled to charge transfer. [Graphical representation proposed by Guidelli (5). Reproduced with permission from G. López Cueto, "Métodos Cinéticos de Electroanálisis," in *Métodos Cinéticos de Análisis*, M. D. Pérez-Bendito and M. Valcárcel, Eds., University of Córdoba, Córdoba (Spain), 1984, p. 211.]

represents an electron transfer at the electrode surface, and C represents a homogeneous chemical reaction. Hence an EC reaction involves a chemical reaction of the product after electron transfer. Chemical species participating in the charge transfer at the electrode are symbolized by O and R, and those participating in only the chemical step as X, Y, or Z.

The CE reaction scheme involves a chemical reaction prior to the charge transfer at the electrode:

$$Z \rightleftharpoons O \qquad\qquad\qquad (C)$$
$$O + ne \rightarrow R \qquad\qquad\qquad (E)$$

The equilibrium between Z and O, which is altered in the reaction layer as O is consumed at the electrode surface, will tend to be re-established by further production of O at the expense of Z. The velocity of the chemical reaction controls the overall process if the former is lower than the velocity of the electron transfer at the electrode yielding a so-called kinetic current. Since the chemical reaction precedes the electrochemical one, the current is also called prekinetic and is characterized by an apparent number of transferred electrons that is smaller than the number exchanged at the electrode.

In the EC reaction mechanism the product of the electrode reaction, R, reacts (e.g., with solvent) producing a species that is not electroactive at potentials where the reduction of O is occurring:

$$O + ne \rightarrow R \qquad\qquad\qquad (E)$$
$$R \rightleftharpoons Y \qquad\qquad\qquad (C)$$

Since the limiting current is controlled by the diffusion of O to the electrode surface, the chemical process does not affect the current but alters the reversibility of the process (producing *postkinetic currents*).

Parallel, catalytic reactions of the EC′ type can be symbolized as follows:

$$O + ne \rightarrow R \qquad\qquad\qquad (E)$$
$$R + Z \rightleftharpoons O + Y \qquad\qquad\qquad (C')$$

The product of the electrode reaction further reacts with another nonelectroactive chemical species (sometimes with itself); this results in a partial or total regeneration of the original species O. The diffusion current is increased because of the regeneration (catalytic cycle) in the vicinity of the electrode. The resulting current is termed the *catalytic current*. Sometimes the catalyst is a nonelectroactive species that lowers the overvoltage for species O by interacting with it:

$$OZ + ne \rightarrow Z + R \qquad\qquad\qquad (E)$$
$$O + Z \rightleftharpoons OZ \qquad\qquad\qquad (C')$$

Catalyst Z in this case affects the reversibility of the electrode process, resulting in a shift of $E_{\frac{1}{2}}$ and the appearance of a "catalytic prewave." The limiting current of this wave is controlled by the rate of the chemical reaction for OZ formation in the reaction layer.

Disproportionation reactions can be illustrated as follows:

$$O + ne \rightarrow R \qquad\qquad\qquad (E)$$

$$R + R \rightleftharpoons O + R' \qquad\qquad (C')$$

The regeneration in this case is only partial, but as in the catalytic mechanism the current is controlled by the rate of the chemical step. The apparent number of electrons exchanged is larger than the real number but cannot increase infinitely as in the catalytic mechanism. Disproportionation currents are sometimes considered to be a particular case of catalytic reactions at the electrode surface.

Alternate reactions of the ECE type can be symbolized in the following manner:

$$O + ne \rightarrow R \qquad\qquad (E)$$
$$R + Z \rightleftharpoons O' \qquad\qquad (C)$$
$$O' + ne \rightarrow R' \qquad\qquad (E)$$

Here a chemical reaction follows a first electrode reaction, and the product of the chemical step undergoes electron transfer at the electrode in a subsequent electrochemical step. The current of the second electrochemical step is controlled by the rate of the chemical step.

Other examples of coupled reactions can be found. From an analytical viewpoint, however, processes affecting the current (i.e., CE and EC' mechanisms) are of particular interest. Of these, catalytic currents resulting in a significant increase over the diffusion current need to be singled out as offering interesting analytical possibilities for determinative purposes because of their low limits of detection and high sensitivity.

### 5.1.2. Analytical Applications of Voltammetric Catalytic Currents

Voltammetric catalytic currents have been known for a long time, but their first analytical application appears to have been contributed by Heyrovsky, the father of polarography. In an attempt to eliminate the interference of cobalt, iron, nickel, and zinc in the polarographic determination of rhenium, he treated solutions of manganese(II) salts with hydrogen sulfide and to his surprise observed an "abnormal" increase and shift in the rhenium wave (7).

He indicated that the catalytic effect was probably due to the electrodeposition of hydrogen catalyzed by a sulfur compound of rhenium and proposed utilizing this effect to determine as little as 0.02 $\mu$g of rhenium in manganese(II) salts.

Calibration curves for catalyst determination are prepared by plotting the catalytic current versus catalyst concentration; this gives linear plots within certain ranges of concentration. Table 5.1 lists some applications of catalytic currents in voltammetry. A discussion of the variables (e.g., catalyst, substrate, ligand–activator, and solution composition) has been offered by Milyavskii (19). Milyavskii's considerations, based on personal experimental studies and other published information, were intended to formulate basic criteria to guide the development of analytical procedures utilizing catalytic polarographic waves.

Additional examples of electrocatalytic determinations based on redox, disproportionation, hydrogen ion discharge, and metal-complex catalytic mechanisms can be found in Refs. 4 and 5.

### 5.2. HETEROGENEOUS CATALYTIC DETERMINATIONS BASED ON IMMOBILIZED ENZYMES

#### 5.2.1. Enzyme Regeneration in Homogeneous Systems

When enzymes are used as analytical reagents for the determination of substrates, their catalytic nature suggests that repetitive use should be possible. This results in better utilization of relatively expensive reagents. Reagent recirculation in closed flow-through systems (20, 21) permits the reutilization of soluble enzyme preparations. Moroever, and in principle, the use of enzymes introduces sufficient selectivity (specificity) that inherently nonspecific methods of detection (e.g., glass pH-sensing electrode, conducto-metric cells, and thermistor-based detection) may be used.

For example, several substrate determinations are based on the general scheme

$$S + H_2O + O_2 \overset{E}{\rightleftharpoons} P + H_2O_2 \tag{5.1}$$

in which S and P are substrate and product, and E is an appropriate enzyme. Determination of substrates such as glucose, uric acid, galactose, and D- and L-amino acids are among those based on the type of reaction represented by eq. 5.1. An example of implementing recycling of the enzyme solution, sample injection into a closed flow-through system, and ampero-

Table 5.1. Examples of Determinations Based on Catalytic Currents

| Species Determined | Electrode[a] | Comments, Reference |
|---|---|---|
| Antimony(III) | Rotating platinum | Enhancement of anodic wave for Br⁻ in acidic media (8). Continuous determination of Sb(III) in effluent streams. |
| Molybdenum(VI) | Rotating graphite disc | Determination of nanogram amounts at pH 2–3. Catalytic wave of $0.10\ M$ bromate develops at $-0.1$ to $-0.7$ V vs. SCE (9). |
| Nitrite | Vitreous graphite and Hg electrodes | Mb(VI) reduced at $-0.48$ V vs. SCE (pH 2.1, 0.1 mM $Na_2MoO_4$, 0.10 $M$ KCl) produces a deposit that selectively catalyzes reduction of nitrite ions (10). In linear potential scan voltammetry the peak current is linear in 0.01–0.10 mM nitrite. No interference from up to 100-fold excess nitrate. |
| Uranium | DME* | Highly selective determination by differential pulse polarography measuring the catalytic peak of nitrate reduction (11). Limit of detection: 1 ppb uranium extracted with 1% $Ph_3AsO$ in $CHCl_3$ prior to determination step. |
| Chromium(VI) | Platinum | Adsorbed iodine acts as catalyst for irreversible electrochemical reduction of Cr(VI) in acidic media (12). |
| Cobalt | Carbon cloth | Detection of cobalt ions in the effluent of ion-exchange chromatographic columns by monitoring the catalytic effect of these ions on the oxidation current of tartrate. Minimum detectable quantity: 5 ng. Application: determination of cobalt in coolants for boiling-water nuclear reactors (13). |

| | | |
|---|---|---|
| Iron(III) | DME | Determination based on catalytic polarographic reduction of bromate in presence of resacetophenone isoniazidhydrazone (14). Application: determination of iron in maize leaves after ashing. |
| Water hardness | DME | Indirect determination based on catalytic wave produced by $Mg^{2+}$ reduction at the DME. Determination depends on the stoichiometric displacement of $Mg^{2+}$ ion from its relatively weak complex with EDTA by the majority of other divalent cations. Proposed method is more rapid than conventional determinations of water hardness (15). |
| Adrenaline | DME | Use of polarographic catalytic current obtained as a result of adrenaline catalysis of the reduction of Ge(IV) in $1 \, M$ $HClO_4$ (16). Application: determination of adrenaline in solutions for hypodermic injection. Detailed study of the electrochemical process involved provided by Mark (17). |
| Cysteine, homocysteine, $N$-acetylcysteine, and glutathione | Carbon paste | Chromatographic detection based on electrocatalytic effect of the analytes on carbon paste electrodes modified by incorporation of cobalt phthalocyanine (18) |

---

[a] SCE = saturated calomel electrode; DME = dropping mercury electrode.

95

metric monitoring of the change in dissolved oxygen level as a result of the overall reaction 5.1, has been offered for glucose determination (21). As already indicated, the advantage of enzyme recirculation is obvious; sample injection with flow-through systems allows processing a large number of samples per unit time, a feature of interest in clinical laboratories with heavy sample demands (22). The overall approach is an example of the analytical use of transient signals generated by series processes and involves signal measurement under dynamic conditions in a system approaching equilibrium; therefore, it qualifies as a kinetic method of determination. The hydrogen peroxide released in reaction 5.1 would interfere with the electrochemical determination of dissolved oxygen. This problem can be circumvented by using glucose oxidase that contains catalase as an impurity, which is inexpensive and insures instantaneous destruction (for all practical purposes) of the hydrogen peroxide:

$$2H_2O_2 \xrightarrow{\text{catalase}} 2H_2O + O_2$$

A diagram of the closed flow-through system is shown in Figure 5.2. Because the amperometric response of the platinum electrode to oxygen is very sensitive to changes in flow rate (23), the flow system is divided in two hydrodynamically independent sections. As a result of difference in elevation between points A and B, the solution flows by gravity at a constant rate from A to B. A peristaltic pump takes the solution back from B to the reservoir. The flow between B and C (0.5 to 0.6 mL/s) is slightly over twice the flow rate between A and B (0.20 to 0.25 mL/s). This aspirates air, bubbling it through the reservoir solution to keep a constant oxygen level responsible for the baseline signal. The reservoir solution, located between branches AB and BC, acts as a pool whose large volume dissipates electrostatic charges generated by the friction of the pump rollers on the plastic tubing transporting the solution.

Figure 5.3 shows typical traces obtained with the setup of Figure 5.2. The baseline corresponds to oxygen saturation; after injection, the glucose sample travels through the coil and reacts with the dissolved oxygen in the circulating reagent solution producing a "plug" in which the oxygen level is lower than it is outside the plug. Because of the predominantly laminar nature of the flow, this plug retains its boundaries (minimum dispersion) and reaches the working electrode that senses the oxygen level in the flowing solution. Restoration to baseline results from the imposed flow and the bubbling of air into the reservoir vessel. The small positive signal preceding the actual peak profile (Figure 5.3) is an artifact resulting from the sudden change of flow produced by the injection. The height of the negative peak provides the analytical information for substrate determination (Figure 5.4). As many as 700

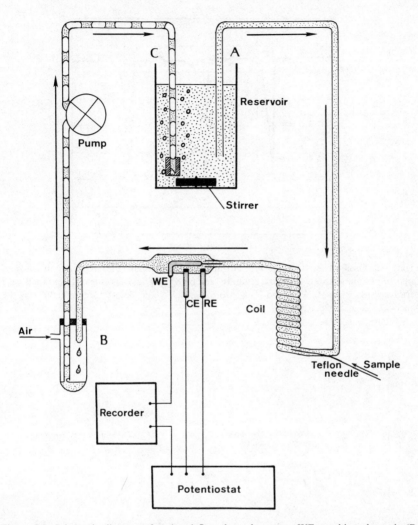

**Figure 5.2.** Schematic diagram of a closed flow-through system. WE, working electrode (Pt wire); CE, counter electrode [Pt/NaCl (*aq*) 7.0 g/L]; RE, reference electrode (SCE). Arrows indicate the direction of flow. Mixing coil: glass tube of 30.0-cm total length, comprising 13 turns with elliptical cross section (1 × 2 mm). [Reprinted with permission from Ch.-M. Wolff and H. A. Mottola, *Anal. Chem.*, **50**, 94. Copyright 1978 by the American Chemical Society.]

measurements per hour (with some precautions even about 1700 measurements per hour) can be performed, and more than 10,000 serum samples (10 $\mu$L each) can be processed with the same reservoir solution (200-mL initial volume).

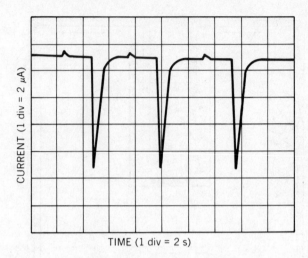

**Figure 5.3.** Oscilloscopic traces of typical transient signals obtained by repetitive injection of three 10-$\mu$L aliquots of a D-glucose solution containing 10 g/L of the sugar. [Reprinted with permission from Ch.-M. Wolff and H. A. Mottola, *Anal. Chem.*, **50**, 94. Copyright 1978 by the American Chemical Society.]

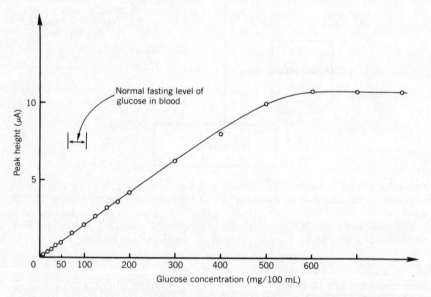

**Figure 5.4.** Working curve for glucose determination. Size of injected sample was 20 $\mu$L. [Reprinted with permission from Ch.-M. Wolff and H. A. Mottola, *Anal. Chem.*, **50**, 94. Copyright 1978 by the American Chemical Society.]

The recycling of both enzyme and coenzyme was demonstrated in the determination of glucose using glucose dehydrogenase (24). Dehydrogenases need nicotinamide adenine dinucleotide (NAD) or its reduced form (NADH) as coenzyme. Recirculation of glucose dehydrogenase, NAD, and glutamate dehydrogenase (for the slow oxidation of NADH back to NAD) and photometric monitoring of NADH (at 340 nm) permitted selective determination of glucose in human blood serum at a rate of 120 samples per h. As many as 300 determinations are possible with the same 40-mL (initial volume) circulating solution and injected sample volumes of 23 $\mu$L (24).

### 5.2.2. Enzyme Regeneration by Use of Immobilized Enzyme Preparations

Glucose oxidase is a relatively inexpensive enzyme that also offers relatively reasonable retention of activity with time in solution. It takes about a month at room temperature for the average activity of a soluble glucose oxidase solution to decrease to 20% of its original value. Storage at $+4°C$ reduces the loss to about 70%.

Some important clinical determinations make use of enzymes that are not sufficiently stable and/or inexpensive to be used in solution for relatively long periods of time and at the high activity levels required for success in closed flow-through systems such as the one in Figure 5.2. For such enzymes immobilization offers a competitive alternative to permit their use as analytical reagents in batch as well as continuous-flow systems. *Immobilization* refers to the localization or confinement of the enzyme molecule in such a manner that it remains physically separated from the substrate solution and the products of the enzyme-catalyzed reaction. An ever-increasing trend in studies using immobilized enzyme preparations, particularly in analytical chemistry, has been documented in the past two decades. This surge in interest results from the advantages that immobilized enzyme preparations offer:

1. Easy separation and recovery from reactants and products of the enzyme-catalyzed reaction (which leads to the possibility of enzyme reuse).
2. Easy use of immobilized enzyme reactors for continuous monitoring of substrate levels (adaptability to continuous-flow analyses),
3. Capability of serving as models for the study of enzymes bound to natural cell membranes of secondary analytical importance.

In general, immobilization can be achieved by chemically or physically attaching the enzyme to a support or by confining the enzyme to a restricted volume by means of a semipermeable membrane. Figure 5.5 graphically

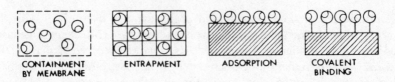

**Figure 5.5.** Graphical illustration of the four most common methods of enzyme immobilization. [From Ref. 30. Reproduced with permission from Academic Press, Inc., New York.]

depicts the main types of immobilization of analytical interest. Containment by membrane and entrapment find use in the construction of *enzyme electrodes*. Physical adsorption has found little use in chemical analysis; probably the only application has been the removal of an enzyme from a closed-loop continuous-flow system for determination of enzyme activity (25). However, covalent binding is extensively used in enzyme reactors for use in continuous-flow systems.

It appears that the first report of enzyme immobilization is due to Nelson and Griffin (26). They observed that invertase (extracted from yeast) was adsorbed on charcoal and that the adsorbed enzyme preparation showed catalytic action like that of the native enzyme. The first successful covalent binding of a variety of proteins (including enzymes) was carried out after World War II. Michael and Evers (27) described the covalent binding of proteins to carboxymethylcellulose in 1949. Enzyme immobilization did not become popular, however, until the 1950s when Grubhofer and Schleith (28, 29) reported coupling carboxypeptidase and amylase to diazotized poly (*p*-aminostyrene). Theoretical and practical aspects of immobilized enzymes have been covered in several reviews (e.g., Ref. 30), and a monograph centered on analytical applications is also available (31).

In chemical immobilization, covalent bond formation between some functional group(s) of the support material and some such group of the enzyme or functional groups of two or more enzyme molecules is involved. The covalent attachment must be, of course, via amino acid residues not essential to the catalytic function of the enzyme. A usual feature of chemical immobilization is irreversibility of the reaction. The original enzyme molecules cannot be separated from the matrix and used as free enzymes again. There are, however, some cases in which the enzyme can be separated from the immobilizing matrix.

The physical immobilization of enzymes depends on entrapment within microcompartments, on containment within special membranes, or on the operation of certain physical interactions such as ionic forces (electrostatic immobilization), hydrogen bonds, or enzyme–enzyme interactions. In theory, physical immobilization should be completely reversible, but there are exceptions.

As heterogeneous catalysts, immobilized enzymes differ in many respects (e.g., optimum pH, Michaelis–Menten constant, and stability) from their soluble counterparts. An exception to this is immobilization by containment of enzyme solutions by membranes. This mild and widely applicable method of immobilization has received little analytical attention, probably because no enhancement of stability can be achieved by this method, a drawback for analytical use of expensive enzymes. Depending on the enzyme, immobilization may result in a remarkable increase of activity retention with time. An example of this is shown in Figure 5.6 for the enzyme uricase covalently bonded to glass.

The two main areas of analytical application of immobilized enzyme preparations are enzyme electrodes and enzyme reactors for use in continuous-flow systems.

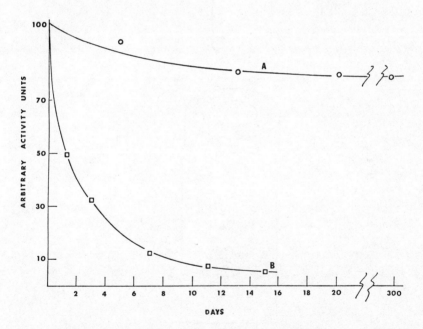

**Figure 5.6.** Operational stability (long term) of uricase (EC 1.7.3.3). (*A*) Enzyme immobilized on controlled pore glass (CPG–3000). (*B*) Soluble preparation. Arbitrary activity units (*y* axis) were normalized to initial activity taken as 100 arbitrary units. Between determinations both preparations were kept at 4° C in sodium borate–ammonium sulfate buffer of pH 9.4. [Reprinted with permission from A. Iob and H. A. Mottola, *Anal. Chem.*, **52**, 2332. Copyright 1980 by the American Chemical Society.]

Most applications of *enzyme electrodes* involve physical entrapment of the enzyme and measurements at steady-state conditions by use of an electrochemical sensor (amperometric or potentiometric) in contact with a layer of the entrapped enzyme. Since enzyme electrodes do not monitor enzyme activity (concentration) and since the electrode per se is not made of enzyme, the denomination "enzyme electrode" for substrate determination is open to criticism. I prefer the name *enzyme reactor*, although I recognize the popular use of the electrode nomenclature. In reporting the performance of an enzyme reactor constructed around a flat-surface pH-sensing electrode and a chamber filled with urease covalently immobilized on nylon shavings (Figure 5.7), competitive analytical performance was indicated by initial rate measurements compared with equilibrium measurements (32). The comparison is illustrated in Table 5.2.

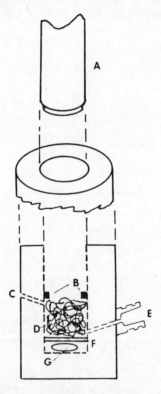

**Figure 5.7.** Details of urease reactor (electrode). A, Flat-bottom combination pH electrode; B, rim to position electrode inside the reactor chamber; C, sample overflow outlet; D, nylon shavings; E, inlet for sample injection; F, perforated circular nylon disk; G, magnetic stirring bar. Shavings and reactor body machined from nylon 6. [From Ref. 32, reproduced with permission from Academic Press.]

**Table 5.2. Comparison of Data of Analytical Interest Extracted from Calibration Plots Applying Equilibrium and Kinetic Information in Determination of Urea with Nylon-Shavings Enzyme Reactor of Figure 5.7**

|  | Equilibrium Determination[a] | Kinetic Determination[b] |
|---|---|---|
| Linear concentration range ($M$) | $1.0 \times 10^{-6}$ to $5.0 \times 10^{-3}$ | $1.0 \times 10^{-6}$ to $1.0 \times 10^{-1}$ |
| Correlation coefficient of calibration plot | 0.999 | 0.998 |
| Sensitivity (slope of calibration plot) | 0.433 pH unit per decade of urea concentration[c] | 0.395 pH unit per decade of urea concentration per min[c] |
|  | 0.375 pH unit per decade of urea concentration[d] | |
| Absolute standard deviation (more than three replicates) | 0.010–0.050 pH units | 0.010–0.060 pH units |

[a] Measurement of equilibrium pH after completion of reaction.
[b] Initial rate in pH units per minute extracted from the first 10% of reaction.
[c] Standards in Tris buffer.
[d] Standards in physiological buffer. Smaller slope may be due to higher ionic strength.
*Source*: Ref. 32, reproduced with permission.

Covalent bonding on the surface of an inert matrix (e.g., glass or nylon) offers the best immobilization approach for preparations to be used in continuous flow (33). This type of immobilization takes advantage of the exposed groups on the enzyme surface. The most commonly used reactions are those involving primary amino groups or the phenol ring of tyrosine. The choice of method depends on (a) the stability of the enzyme at the pH at which coupling is performed, (b) the stability of the linkage at the pH at which the immobilized enzyme preparation will be used, and (c) the carrier. Background information on the reactions involved can be found in monographs by Means and Feeney (34) and by Zaborsky (35). Among analytical applications, one of the most widely used methods involves alkylamino glass and glutaraldehyde. Although the detailed chemistry of this reaction is not well understood, the individual steps include the reaction of the silica framework with an aminosilane, modification of the product of this reaction with the difunctional reagent glutaraldehyde, and finally immobilization of the enzyme. These steps can be symbolized as follows:

$$-O-Si-OH + (C_2H_5O)_3Si(CH_2)_3NH_2 \rightarrow -O-Si-O-Si(CH_2)_3NH_2 \quad (1)$$

$$-NH_2 + OHC-(CH_2)_3-CHO \rightarrow -N=CH-(CH_2)_3-CHO \quad (2)$$

$$-CHO + H_2N-E \rightarrow -CH=N-E \quad (3)$$

where $H_2N-E$ = enzyme.

Immobilization on glass through glutaraldehyde attachment is attractive because of (a) its simplicity and mild operating conditions, (b) its success with quite a variety of enzymes, and (c) the good mechanical and chemical stability of glass. Several types of reactor configurations can be adapted to continuous-flow operations. The most common ones are: (a) packed columns, (b) open-tube wall reactors, and (c) single-bead-string reactors.

Most commonly reported in the literature are packed-column reactors, with enzyme preparations made using the glutaraldehyde linkage and controlled-pore glass as the inert matrix. Analytical aspects of packed-column immobiliz-ed-enzyme reactors with emphasis on thermochemical detection have been discussed by Schifreen et al. (36). Their observations indicate that the rate of product formation is controlled by mass transfer of the substrate to the surface of the carrier or to diffusion within the porous glass rather than by the rate of the enzyme-catalyzed reaction. They also indicate that to a first approxim-ation a packed-bed reactor can be modeled in terms of reaction plates corresponding to separation plates in chromatographic columns. The same mathematics that apply to chromatographic columns apply then to packed-column enzyme reactors (36).

A typical example of packed-column enzyme reactors in continuous-flow systems is shown in Figure 5.8, adapted from Masoom and Townshend (37). It illustrates manifold configurations for the simultaneous as well as sequential determination of two species (sucrose and glucose) by use of two packed-column enzyme reactors and amperometric detection of the hydrogen peroxide produced in the following reactions:

$$\text{Sucrose} \xrightarrow{\text{invertase, mutarotase}} \beta\text{-D-glucose}$$

$$\beta\text{-D-glucose} + H_2O + O_2 \xrightarrow{\text{glucose oxidase}} \beta\text{-D-gluconic acid} + H_2O_2$$

Typical signal profiles for the simultaneous determination of sucrose and glucose in soft drinks are shown in Figure 5.9.

The use of open tubes with an enzyme immobilized at the inner wall seems to

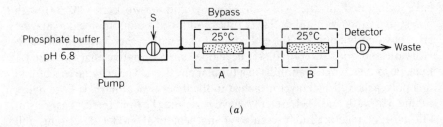

(a)

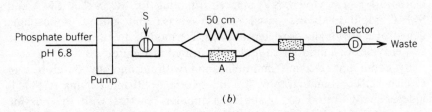

(b)

**Figure 5.8.** Flow injection manifolds for the determination of sucrose and glucose. (a) Manifold for sequential determination. (b) Manifold for simultaneous determination. Column A contains immobilized invertase and mutarotase. Column B contains immobilized glucose oxidase. S = sample injection port. [From reference 37, reproduced with permission.]

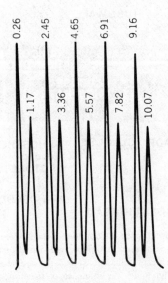

**Figure 5.9.** Typical signal profiles for the simultaneous determination of sucrose and glucose in soft drinks. Numbers indicate times at which peaks appear, in minutes. [From Ref. 37, reproduced with permission.]

have originated in the work of Hornby and co-workers (38, 39) with polystyrene and nylon tubes. These tubes contained a monomolecular enzyme layer covalently bound to the etched inner wall with rather low local activity. Horvath and co-workers (40, 41) developed an alternative that provided a thick porous enzymatic annulus in a tubular envelope. The enzyme was bound to a polycarboxylic gel layer attached to the inner wall of narrow-bore nylon tubing. Horvath and Pedersen (42) have reviewed the fundamental aspects of this type of reactor and presented a mathematical model describing coil performance in air-segmented continuous-flow systems. The Technicon Corporation (Tarrytown, NY) markets a module for use in their SMA and SMAC rapid analyzers based on hexokinase and glucose-6-phosphate dehydrogenase for glucose determination. The module uses a gel layer with the two enzymes co-immobilized and is illustrated in Figure 5.10.

By combining enzyme immobilization with technology developed in capillary gas chromatography, Iob and Mottola (43) and Kojima et al. (44) independently introduced glass open tubular reactors with the enzyme chemically bonded to the inner wall of narrow-bore glass tubing. To increase

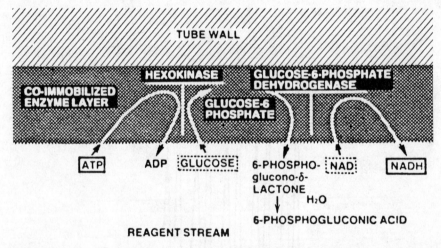

**Figure 5.10.** Schematic representation of the bound-enzyme hexokinase method for determination of glucose. The hexokinase and glucose-6-phosphate dehydrogenase (G6PDH) are coated on the inner wall of a disposable plastic coil. Glucose from the sample and the reagent ATP diffuse into the enzyme layer and react under the influence of the bound hexokinase to form the intermediate product glucose-6-phosphate. The latter reacts with G6PDH to produce 6-phosphoglucono-δ-lactone, with concomitant formation of NADH from NAD. The amount of NADH formed is proportional to the glucose concentration and can be monitored at 340 nm.

the surface area available for immobilization, silica filament "whiskers" were grown by adaptation of an ammonium hydrogen fluoride treatment (45). Typical applications of such reactors are the determination of uric acid by means of immobilized uricase and amperometric monitoring of dissolved oxygen levels (43) and L-lactate with immobilized lactate dehydrogenase and photometric monitoring of NADH (46).

Although whisker growth increases the surface area by about three orders of magnitude, their preparation is elaborate, involves the use of corrosive chemicals, and frequently results in nonuniform surface coverage as a result of imprecise temperature control. In the quest for a simpler mode of increasing surface area of open tubular reactors that would also result in more uniform surface coverage, Gosnell et al. (47) proposed the preparation of open tubular reactors by thermally embedding controlled-pore glass chips on the walls of plastic tubing. The higher local activity of these reactors when compared with whisker-modified reactors (Figure 5.11) is of analytical interest; their simplicity and fast preparation is of additional appeal.

Penicillinase (EC 3.5.2.6) immobilized on glass beads by glutaraldehyde attachment after ammonium hydrogen fluoride etching was made part of a single-bead-string reactor used for the determination of penicillins in pharmaceutical tablets, injectable drugs, and fermentation broths (48). Hwang and Dasgupta (49) developed an unsegmented continuous-flow system for the determination of sub-part-per-billion levels of aqueous hydrogen peroxide that makes use of horseradish peroxidase chemically immobilized on an aminated microporous polymeric adsorbent and in a single-bead-string reactor configuration. The determination is based on the peroxidase-catalyzed oxidation of nonfluorescent p-hydroxyphenylacetate to its fluorescent dimer. A detection limit of 0.3 ppb and a large dynamic range (1 ppb to 1 ppm) are reported by the authors and are attractive features of the proposed methodology. Organic peroxides, however, inhibit enzyme activity, apparently in an irreversible manner.

A combination of controlled-pore glass embedded on plastic tubing with etched glass beads in a single-bead-string reactor may well be the optimum configuration for immobilized-enzyme reactors to use in continuous-flow sample processing.

Batch determinations using immobilized enzymes do not seem to have been reported except for the use of enzyme electrodes. Closer to this type of application seems to be the innovation of a tubular reactor in the observation cell of a stopped-flow system (50). This configuration, making use of enzymes immobilized on the walls of nylon tubing, has been used to determine glucose and lactate (51). In this type of reactor arrangement, the enzyme-catalyzed reaction occurs under static conditions and the kinetics is controlled by diffusion and the nature of the reactants in question.

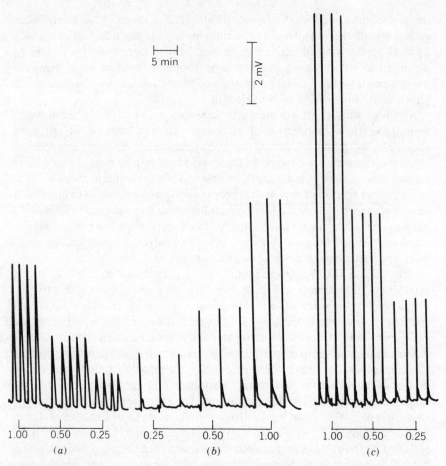

**Figure 5.11.** Typical flow injection signal profiles for injection of penicillin G into 1 mM phosphate buffer carrier of pH 6.40. (a) 3.6-m borosilicate glass open tubular reactor. (b) 0.35-m CPG-Tygon reactor (CPG of 75-Å nominal pore diameter). (c) 0.50-m CPG-Tygon reactor (CPG of 3000-Å nominal pore diameter). CPG = controlled-pore glass. [Reprinted with permission from M. C. Gosnell et al., *Anal. Chem.*, **58**, 1585. Copyright 1986 by the American Chemical Society.]

## REFERENCES

1. V. V. S. Eswara Dutt and H. A. Mottola, *Anal. Chem.*, **48**, 80 (1976).

2. C. Sánchez-Pedreño, M. Hernández Córdoba, and G. Martinez Tudela, *An. Quim. Ser. B*, **77**, 68 (1981).

3. T. Erdey-Gruz, *Kinetics of Electrode Processes*, Wiley-Interscience, New York, 1972, p. 17.

4. H. B. Mark and G. A. Rechnitz, *Kinetics in Analytical Chemistry*, Wiley-Interscience, New York, 1968, pp. 56–60.

5. R. Guidelli, in *Electroanalytical Chemistry*, Vol. 5, A. J. Bard, Ed., Dekker, N.Y., 1971, pp. 250–284.

6. A. J. Bard and L. R. Faulkner, *Electrochemical Methods: Fundamentals and Applications*, Wiley, New York 1980, Chapter 11.

7. J. Heyrovsky, *Nature*, **135**, 870 (1935).

8. L. R. Taylor, R. J. Davenport, and D. C. Johnson, *Talanta*, **20**, 947 (1973).

9. V. F. Toropova, V. A. Vekslina, and N. G. Chovnik, *Zh. Anal. Khim.*, **28**, 967 (1973); *Chem. Abstr.*, **79**, 73203j (1973).

10. J. A. Cox and A. F. Brajter, *Anal. Chem.*, **51**, 2330 (1979).

11. R. Keil, *Fresenius' Z. Anal. Chem.*, **292**, 13 (1978).

12. J. H. Larochelle and D. C. Johnson, *Anal. Chem.*, **50**, 240 (1978).

13. Y. Takata, F. Mizuniwa, and C. Maekoya, *Anal. Chem.*, **51**, 2337 (1979).

14. K. Murali Mohan and S. Brahmaji Rao, *Talanta*, **27**, 905 (1980).

15. M. C. Cheney, D. J. Curran, and K. S. Fletcher III, *Anal. Chem.*, **52**, 942 (1980).

16. J. M. López Fonseca and María del Carmen Arrendondo, *Analyst (London)*, **108**, 847 (1983).

17. H. B. Mark, Jr., *J. Electroanal. Chem.*, **7**, 276 (1964).

18. M. K. Halbert and R. P. Baldwin, *Anal. Chem.*, **57**, 591 (1985).

19. Yu. S. Milyavskii, *Zh. Anal. Khim.*, **34**, 1669 (1979); *J. Anal. Chem. USSR (Engl. Transl.)*, **34**, 1293 (1979).

20. H. U. Bergmeyer and A. Hagen, *Fresenius' Z. Anal. Chem.*, **261**, 333 (1972).

21. Ch.-M. Wolff and H. A. Mottola, *Anal. Chem.*, **50**, 94 (1978).

22. L. T. Skeggs, Jr., *Am. J. Clin. Pathol.*, **28**, 311 (1957).

23. Ch.-M. Wolff and H. A. Mottola, *Anal. Chem.*, **49**, 2118 (1977).

24. P. Roehrig, Ch.-M. Wolff, and J. P. Schwing, *Anal. Chim. Acta*, **153**, 181 (1983).

25. S. M. Ramasamy, A. Iob, and H. A. Mottola, *Anal. Chem.*, **51**, 1637 (1979).

26. J. M. Nelson and E. G. Griffin, *J. Am. Chem. Soc.*, **38**, 1109 (1916).

27. F. Michael and J. Evers, *Makromol. Chem.*, **3**, 200 (1949).

28. N. Grubhofer and L. Schleith, *Naturwissenschaften*, **40**, 508 (1954).

29. N. Grubhofer and L. Schleith, *Z. Physiol. Chem.*, **297**, 108 (1954).

30. J. Konecny, in *Survey of Progress in Chemistry*, Vol. 8, Academic Press, New York, 1977, pp. 195–251.

31. P. W. Carr and L. D. Bowers, *Immobilized Enzymes in Analytical and Clinical Chemistry*, Wiley, New York, 1980.

32. K. D. Begum and H. A. Mottola, *Anal. Biochem.*, **142**, 1 (1984).

33. H. A. Mottola, *Anal. Chim. Acta*, **145**, 27 (1983).

34. G. E. Means, and R. E. Feeney, *Chemical Modification of Proteins*, Holden Day, San Francisco, 1971.

35. O. R. Zaborsky, *Immobilized Enzymes*, CRC Press, Cleveland, Ohio, 1973.

36. R. A. Schifreen, D. A. Hanna, L. D. Bowers, and P. W. Carr, *Anal. Chem.*, **49**, 1929 (1977).

37. M. Masoom and A. Townshend, *Anal. Chim. Acta*, **171**, 185 (1985).

38. W. E. Hornby, D. J. Inman, and A. Mcdonald, *FEBS Lett.*, **9**, 8 (1970).

39. D. J. Inman and W. E. Hornby, *Biochem. J.*, **129**, 225 (1972).

40. C. Horvath and B. Solomon, *Biotechnol. Bioeng.*, **14**, 885 (1972).

41. C. Horvath, A. Sardi, and J. S. Woods, *J. Appl. Physiol.*, **34**, 181 (1973).

42. C. Horvath and H. Pedersen, in *Advances in Automated Analysis*, 7th Technicon International Congress, December 1976, New York City. Technicon Instrument Corporation, Tarrytown, N.Y., 1977.

43. A. Iob and H. A. Mottola, *Clin. Chem. (Winston–Salem, N. C.)*, **27**, 195 (1981).

44. T. Kojima, Y. Hara, and F. Morishita, *Bunseki Kagaku*, **32**, E101 (1983)

45. F. I. Onusca, M. E. Comba, T. Bistricki, and J. Wilkinson, *J. Chromatogr.*, **142**, 117 (1977).

46. F. Morishita, Y. Hara, and T. Kojima, *Bunseki Kagaku*, **33**, 642 (1984).

47. M. C. Gosnell, R. E. Snelling, and H. A. Mottola, *Anal. Chem.*, **58**, 1585 (1986).

48. R. Gnanasekaran and H. A. Mottola, *Anal. Chem.*, **57**, 1005 (1985).

49. H. Hwang and P. K. Dasgupta, *Mikrochim. Acta*, 77 (1985).

50. R. Q. Thompson and S. R. Crouch, *Anal. Chim. Acta*, **144**, 155 (1982).

51. R. Q. Thompson and S. R. Crouch, *Anal. Chim. Acta*, **159**, 337 (1984).

# RATE DETERMINATIONS USING
# UNCATALYZED REACTIONS

Although the bulk of kinetic-based determinations depend on catalytic systems, some analytical applications using the rate of uncatalyzed reactions are of analytical interest, mainly, those involving organic chemical species. If a suitable equilibrium method is available for a given chemical species, a rate measurement approach based on an uncatalyzed reaction will probably not be considered by practicing analysts. The limited applicability of rates of uncatalyzed reactions, however, should not prevent their consideration in this monograph since in particular instances they may solve analytical problems. The considerations that follow center on determinations of single species by uncatalyzed reactions; the use of such reactions for the simultaneous determination of two or more species is covered in Chapter 7.

## 6.1. RATE CONSIDERATIONS FOR THE DETERMINATION OF A SINGLE SPECIES IN UNCATALYZED PROCESSES

From a practical analytical viewpoint, methods using uncatalyzed reaction rates are limited to those following pseudo-zero-order, first-order, and pseudo-first-order kinetics, since most of the chemical systems used involve bimolecular reactions, and orders higher than two are rare and of no analytical practicality.

### 6.1.1. Pseudo-Zero-Order Conditions (Initial Rate Method)

Consider the most common case, a bimolecular reaction of the type

$$A + R \underset{k_b}{\overset{k_f}{\rightleftharpoons}} P$$

where $A$ = species of analytical interest (analyte), $R$ = added reagent, $P$ = product(s), $k_f$ = forward rate coefficient, and $k_b$ = backward rate

111

coefficient. The general differential rate expression is

$$-\frac{d[A]_t}{dt} = -\frac{d[R]_t}{dt} = \frac{d[P]_t}{dt} = k_f[A]_t[R]_t - k_b[P]_t \qquad (6.1)$$

As discussed in Chapter 1, if the rate data are taken in the first 1 to 2% completion of the total reaction, the concentration of reactants can be considered constant and equal to the respective initial concentrations. This allows us to ignore contributions from the reverse reaction. Equation 6.1 then simplifies to a pseudo-zero-order expression:

$$\frac{d[P]_t}{dt} = k_f[A]_0[R]_0 \simeq \text{constant} \qquad (6.2)$$

Equation 6.2 provides the basis for initial rate measurement since, if $[R]_0 = \text{constant}$, the initial rate plotted against $[A]_0$ should be a straight line useful as a calibration plot for the determination of $[A]_0$.

### 6.1.2.  First-Order and Pseudo-First-Order Conditions

A first-order irreversible reaction can be written as

$$A \xrightarrow{\quad k_A \quad} x\,P$$

where $k_A$ = rate coefficient, and $x$ = number describing the stoichiometry of the reaction. The rate expression can be written as

$$-\frac{d[A]_t}{dt} = k_A[A]_t \qquad (6.3)$$

where $[A]_t$ = concentration of A at any time $t$.

Integrating eq. 6.3 provides a relationship between $[A]_t$ and the initial concentration, $[A]_0$, the sought for information:

$$[A]_t = [A]_0\,e^{-k_A t} \qquad (6.4)$$

Substituting eq. 6.4 into eq. 6.3 gives

$$-\frac{d[A]_t}{dt} = k_A[A]_0\,e^{-k_A t} \qquad (6.5)$$

Equation 6.5 is the basis for the *derivative approach* to rate-based determination of species A. Equation 6.4 is the basis for the two different *integral approaches* to kinetic-based determinations: the *fixed-time* and the *variable-time* procedures. If the main reaction is run under pseudo-first-order conditions (where one of the reactants is present in large excess) and if the reverse reaction can be neglected (because the large excess of one of the reactants drives the reaction to virtual completion), then the rate expression takes the form (R in excess).

$$-\frac{d[A]_t}{dt} \simeq k_f[R]_0[A]_t = k'_f[A]_t \qquad (6.6)$$

Equation 6.6 permits pseudo-first-order reactions to be mathematically treated as true-order reactions. A very common procedure in analytical applications is to select experimental conditions to achieve pseudo-first-order conditions.

## 6.2. SPECIAL CASES FOR THE DETERMINATION OF A SINGLE SPECIES IN A MIXTURE BY USE OF AN UNCATALYZED REACTION

If different chemical species present in a mixture have significant different rates of reaction with a common reagent, it may be possible to treat each species separately. That is, during a given time interval only one of the species may be considered as reacting because the other species have already reacted or react so slowly that they do not interfere in the rate measurements. The establishment of conditions under which one of the components can be determined in a mixture undergoing reaction without a significant error being introduced by the presence of other components is then not difficult (1). The criteria for neglecting reactions of *slower-reacting components* or *faster-reacting components* are the ratio of the corresponding rate coefficients, the time selected for measurement, and the errors tolerable for the particular case considered. Figure 6.1 shows the percentage error in the determination of a chemical species A resulting from the interference of a slower-reacting species B and considering $[A]_0 = [B]_0$ as a function of the ratio of coefficients, $k_A/k_B$, and the elapsed time before measurement (represented by the percent of A reacted). If the model reactions of Figure 6.1 are kinetically first-order with respect to both A and B (i.e., for all practical purposes $[R]_0 = [R]_t$ since R is in large excess), the rates of disappearance of A and B are given by expressions similar to eq. 6.3, which in integrated form become.

$$\ln[A]_t = \ln[A]_0 - k_A t \qquad (6.7)$$

$$\ln[B]_t = \ln[B]_0 - k_B t \qquad (6.8)$$

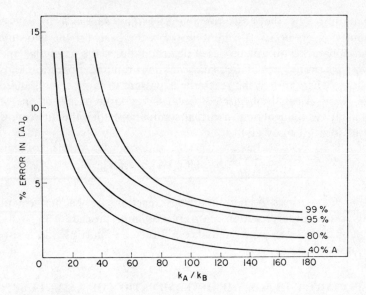

**Figure 6.1** Percentage error in $[A]_0$ resulting from the amount of slower-reacting component B reacting in the time required for a fixed percentage of component A to react plotted as a function of the ratio of rate coefficients. First-order reactions of the type

$$A + R \xrightarrow{k_A} \text{Products}$$

$$B + R \xrightarrow{k_B} \text{Products}$$

are assumed. [From Ref. 1 reprinted with permission.]

If component A is the faster-reacting one, the ratio $k_A/k_B$ can be calculated from eqs. 6.7 and 6.8 to give

$$\frac{k_A}{k_B} = \frac{\ln([A]_0/[A]_t)}{\ln([B]_0/[B]_t)}$$

For 99% reaction of A and 1% reaction of B, this results in

$$\frac{k_A}{k_B} = \frac{\log 100}{\log 1.01} = 463$$

This indicates that if the ratio of rate coefficients is at least $463:1$, it is possible to find a reaction time such that the faster reaction is within 1% of completion while the slower reaction has proceeded to no more than 1%. This is actually a special case of the differential reaction approach discussed in a later chapter.

Efforts to elucidate the relationship between the reactivity and the structure of organic compounds have provided a wealth of rate data of analytical interest (2, 3) that can be used to guide analytical method development.

## 6.3. SELECTED ANALYTICAL APPLICATIONS OF RATES OF UNCATALYZED REACTIONS

### 6.3.1. Determination of Inorganic Species

Although, as already mentioned, the determination of organic species has received most benefit from the utilization of uncatalyzed reaction rates, a few methods have been proposed for the determination of inorganic species. These applications have resulted in part from exploiting ligand-exchange reactions and from using fluorimetric monitoring, which permits achieving low limits of detection and relatively large dynamic ranges.

The number of kinetic-based methods for determining alkaline earth ions is very limited because these ions do not participate in catalytic cycles. A selective determination of magnesium has been proposed based on the reaction, in alkaline medium, with 2-fluoroaldehyde-2-pyridylhydrazone (FAPH) (4). At high pH values (e.g., 12.5–13.5) micellar magnesium hydroxide forms an intense green fluorescent species with FAPH. The fluorescence decreases with time, and its monitoring at 520 nm allows preparation of calibration graphs of initial rate versus magnesium concentration in the range of 0.36 to 1.22 $\mu$g/mL. Determination is possible in the presence of 10-fold concentrations of beryllium and calcium and 25-fold concentrations of strontium and barium; but only 0.5 $\mu$g/mL of mercury(II), iron(III), chromium, and aluminum and 0.4 $\mu$g/mL of EDTA can be tolerated.

Aluminum is another example of a metallic ion species that eludes catalytic determinations. Aluminum in the 0.020-to-10.0 $\mu M$ range can be determined by an initial rate fluorometric method (5). Fluorescence of its chelate with 2-hydroxy-1-naphthaldehyde $p$-methoxybenzoylhydrazone provides'the basis for such a determination.

Table 6.1 lists other selected applications published relatively recently.

### 6.3.2. Determination of Organic Species

The kinetics of uncatalyzed peroxydisulfate oxidation of organic materials in fresh water has been examined by Goulden and Anthony (18). Such oxidation

**Table 6.1. Selected Determinations of Inorganic Species by Means of Rates of Uncatalyzed Reactions**

| Species Determined | Comments | Reference |
|---|---|---|
| Cerium | Oxidation [by Ce(IV)] of 4,8-diamino-1,5-dihydroxyanthraquinone-2,6-disulfonate. Initial rate fluorometric measurement. | 6 |
| Copper | Ligand exchange between the Cu(II) complex with 2-(2-thiazolylazo)-5-dimethylaminophenol and EDTA. Stopped-flow mixing and photometric monitoring (563 nm). | 7 |
| Gallium | Substitution of Cu(II) by Ga(III) in the copper–EDTA complex. Photometric monitoring (240 nm). Application to the determination of gallium in bauxite. | 8 |
| Germanium | Determination as germanate by formation of molybdenum germanate blue. Several species interfere. | 9 |
| Iron | Oxidation [by Fe(III)] of 1,4-diamino-2,3-dihydroxyanthraquinone. Fluorometric monitoring. Thallium(III) also determined. | 10 |
| Manganese | Method of tangents by photometric or fluorometric monitoring of the manganese (probably as $MnO_2 \cdot H_2O$) oxidation of 4,8-diamino-1,5-dihydroxyanthraquinone-2,6-disulfonate. Method applied to the determination of manganese in fumes of industrial workplaces. | 11 |
| Nitrite | Determination in water samples by stopped-flow mixing of the diazotization reaction of sulfanilamide. The diazotization product is coupled with 1-naphthylethylenediamine to yield a highly colored azo dye ($\lambda_{max}= 540$ nm). | 12 |
| Perbromate | Method of tangents. Spectrophotometric monitoring of the $I_3^-$–perbromate reaction. | 13 |
| Phosphorus | Determination based on the phosphomolybdenum blue reaction (reduction of molybdophosphoric acid by ascorbic acid). Determination in check grain and feed samples. | 14 |
| Selenium | As selenite. Turbidimetric fixed-time procedure based on reduction to elemental selenium by ascorbic acid. Determination at the $10^{-5} M$ level by photometric monitoring (460 nm). | 15 |

**Table 6.1** (*continued*)

| Species Determined | Comments | References |
|---|---|---|
| Thorium | Determination in uranium samples by photometric monitoring of the rate of exchange between the Th complex with DCTA and Arsenazo III as attacking ligand. | 16 |
| Vanadium | Oxidation of 1-amino-4-hydroxyanthraquinone and fluorescence measurement. Initial rate determination of vanadium in crude petroleums. | 17 |

in natural-water samples and of single organic compounds can be represented by simple rate equations. This study suggests the use of the uncatalyzed peroxydisulfate oxidation as a convenient route to an automated, wet-chemical system for the determination of organic carbon in some water samples. The usefulness of rate information coupled with determinations of chemical oxygen demand (COD) has also been discussed by Goodwin et al. (19). They presented reaction rate data for 30 model compounds grouped into four categories according to extent of reaction after 120 min. Goodwin et al. also proposed a multiparameter system (including rate data) that reflects changes in both the quantity and overall nature of the oxidizable material. The study was extended to include samples from reactor effluents and from a sewage plant.

Fifteen different phenol species in methanol and acetic acid solutions were individually determined by a fixed-time procedure based on the measurement of the yellow products resulting from the oxidation of the phenols with metaperiodate (20). The reactions were run at 50°C for time periods ranging from 15 min to 2 h, depending on the species being determined. Interestingly, attempts to apply the approach for differential reaction rate determination in mixtures were unsuccessful.

The concentration level of the enzymatically active form of pyridoxal 5'-phosphate can indicate vitamin $B_6$ group deficiency. Hassan and Rechnitz (21) proposed a simple, selective, and reliable method for the determination of this phosphate by monitoring the rate of ammonia liberation (with an ammonia-gas potentiometric sensor) from excess L-tryptophan by action of tryptophanase apoenzyme as a function of concentration of pyridoxal 5'-phosphate coenzyme. The procedure requires less time and affords a better limit of detection ($10^{-8}$ $M$ level) than other proposed methods. No interference was observed in the presence of 100-fold amounts of other members of the vitamin $B_6$ group and their phosphate derivatives.

The normal biuret procedure for total protein determination (which is equilibrium-based) requires 15 to 30 min for full color development. A two-

**Table 6.2. Selected Determinations of Organic Species by Means of Rates of Uncatalyzed Reactions**

| Species Determined | Comments | Reference |
|---|---|---|
| Formaldehyde and hexamethylenetetramine | Monitoring of rate of $CN^-$ addition with a cyanide-selective electrode in a variable-time procedure. Determination of hexamethylenetetramine in pharmaceutical preparations. | 25 |
| Aliphatic amines (e.g., butylamine, isobutylamine, trioctylamine) | Determination in wastewater by reaction with tetrachloro-p-benzoquinone in a variety of solvents. | 26 |
| Thiamine | Oxidation by Hg(II) in basic solution with formation of a fluorescent thiochrome. Initial rate measurements. Determination of thiamine in vitamin and mineral preparation. | 27 |
| Creatinine | Determination in serum based on the Jaffé reaction. Monitoring with a picrate-selective electrode. Fixed-time (30 and 270 s). Creatinine adsorbed (from acidified serum) on strong cation exchange resins. Kinetic determination applied to eluted creatinine. Method free from optical interferences and from bilirubin interference. | 28 |
| Tetracycline and oxytetracycline | Degradation rates (in strong acidic medium) to the anhydro derivatives followed spectrophotometrically (decrease of absorbance at 393 nm due to parent compound). Determination in powders and capsules of pharmaceutical origin. | 29 |
| Sulfonamides | Variable-time procedure using stopped-flow mixing and photometric monitoring of the color developed in the reaction with nitrite and 1-naphthylethylenediamine (the Griess reaction). Application to the analysis of urine samples. | 30 |
| Amino acids (alanine, phenylalanine, leucine, isoleucine) | Reaction with ninhydrin. Multipoint kinetics and nonlinear regression analysis. Absorbance versus time data (1 to 3 half-lives with reaction pseudo-first-order in amino acid) fitted to first-order model to predict absorbance change at equilibrium. Results practically independent of temperature changes and of ninhydrin concentration. | 31 |

118

| Analyte | Description | Ref. |
|---|---|---|
| Serum low-density lipoprotein | Fixed-time turbidimetric determination. Turbidity results from specific precipitation of the lipoprotein by heparin, calcium ions, EDTA, and lipase. | 32 |
| Ethanol | Oxidation by permanganate and photometric monitoring of $MnO_4^{2-}$ (400 nm). | 33 |
| Ascorbic acid | Based on the linear relation, over a wide range of concentration, between the apparent first-order rate coefficient for the reduction of 2,6-dichloroindophenol and the concentration of ascorbic acid. Highly selective method that allows determination in turbid or colored food samples. | 34 |
| Uric acid | Determination in whole (undeproteinized) serum based on the rate of reduction of iron(III) in the presence of 2,4,6-tripyridyl-s-triazine. Fixed-time measurements afford selective determinations. | 35 |
| Kojic acid | Stopped-flow mixing and absorptiometric monitoring at 525 nm of the reaction with 2,6-dichlorophenolindophenol. Observed second-order rate coefficient proportional to the kojic acid concentration. Application to the determination of this acid in fermentation media. | 36 |
| Acetaminophen | Determination in pharmaceutical tablets by measurements of its rate of bromination. | 37 |
| Tryptophan | Kinetic-fluorometric method (fixed-time). Reaction with formaldehyde at a pH of 10.8 (carbonate buffer). Determination of tryptophan in foods and foodstuffs. | 38 |
| Coricosteroids | Reaction rate method based on a modification of the blue tetrazolium reaction. The red formazan formed is monitored (fixed-time) at 525 nm. Determination in pharmaceutical skin preparations. | 39 |
| 1-Piperidinocyclohexanecarbonitrile (PCC) | Detection and semiquantitative determination of PCC in illicit samples by a fixed-time procedure based on reaction with p-nitrobenzaldehyde and o-dinitrobenzene. Absorbance of quenched solutions measured at 555 nm after 45-min reaction. | 40 |

point, fixed-time rate method based on the same chemistry with the aid of stopped-flow mixing allows determinations in 10 s with good precision (22). By sacrificing some precision, determinations can be made in 0.2 s.

A kinetic, accelerated-oxidation method has been applied to the study of the natural antioxidant content changes in sea buckthorn (*Hippophae*) oil (23). The method was applied to evaluate the stability of the oil. Kinetic curves of cumene oxidation in the presence of the oil samples indicated three types of antioxidants: $\alpha$-tocopherols, $\gamma$-tocopherols, and some other antioxidants, both strong and mild.

Organic species that react with iodine and bromine have been determined ($10^{-6}$ to $10^{-4}$ $M$ range) by in situ chemical generation of bromine or iodine ($H_2O_2$ oxidation of the corresponding halides in acidic medium) (24). 1,10-Phenanthroline or $o$-toluidine were used as indicators that react more slowly with $Br_2$ or $I_2$ than the sought-for species. Species typically determined by measurement of the time delay (pseudoinduction period) are thiosalicylic acid, hydroquinone, thiuram E, phenylhydrazine, thiourea, 1-naphthol, phenol, and ascorbic acid.

Table 6.2 lists additional determinations of organic species by rate measurement of uncatalyzed chemical reactions.

## REFERENCES

1. H. B. Mark, Jr. and G. A. Rechnitz, *Kinetics in Analytical Chemistry*, Wiley, New York, 1968, pp. 72–75.

2. L. P. Hammett, *Physical Organic Chemistry*, 2nd ed., McGraw-Hill, New York, 1970.

3. M. Kopanica and V. Stara, in *Wilson and Wilson's Comprehensive Analytical Chemistry*, Vol. 28, G. Svehla, Ed., Elsevier, Amsterdam, 1983, pp. 76–81.

4. J. J. Laserna, A. Navas, and F. García Sánchez, *Microchem. J.*, **27**, 312 (1982).

5. P. C. Ioannou and P. A. Siskos, *Talanta*, **31**, 253 (1984).

6. A. Navas, F. Sánchez Rojas, and F. García Sánchez, *Mikrochim. Acta*, **1**, 175 (1982).

7. K. Nakagawa, T. Ogata, K. Haraguchi, and S. Ito, *Bunseki Kagaku*, **29**, 319 (1980).

8. T. Nozaki and M. Sakamoto, *Bunseki Kagaku*, **30**, 196 (1981).

9. V. K. Rudenko and N. N. Adamenko, *Ukr. Khim. Zh.*, **46**, 861 (1980); *Chem. Abstr.*, **93**, 160641h (1980).

10. F. Salinas, C. Genestar, and F. Grases, *Anal. Chim. Acta*, **130**, 337 (1981).

11. A. Navas and F. Sánchez Rojas, *Talanta*, **31**, 437 (1984).

12. M. A. Koupparis, K. M. Walczak, and H. V. Malmstadt, *Analyst (London)*, **107**, 1309 (1982).

13. L. A. Lazarou, P. A. Siskos, M. A. Koupparis, T. P. Hadjiioannou, and E. H. Appelman, *Anal. Chim. Acta*, **94**, 475 (1977).

14. M. S. McCracken and H. V. Malmstadt, *Talanta*, **26**, 467 (1979).

15. T. E. Gaytan, E. Alba, and Yu. A. Zhukov, *CENTRO, Ser. Quim. Tecnol. Quim.*, **5**, 59 (1977); *Chem. Abstr.*, **92**, 14772q (1980).

16. V. M. Aleksandruk, M. A. Nemtsova, and A. V. Stepanov, *Radiokhimiya*, **23**, 778 (1981); *Chem. Abstr.*, **96**, 45502p (1982).

17. F. Salinas, F. García Sánchez, F. Grases, and C. Genestar, *Anal. Lett.*, **13** (A6), 473 (1980).

18. P. D. Goulden and D. H. J. Anthony, *Anal. Chem.*, **50**, 953 (1978).

19. A. E. Goodwin, D. K. Cabbiness, and H. A. Mottola, *Water, Air, Soil Pollut*, **8**, 467 (1977).

20. L. R. Sherman, V. L. Trust, and H. Hoang, *Talanta*, **28**, 408 (1981).

21. S. S. M. Hassan and G. A. Rechnitz, *Anal. Chim. Acta*, **126**, 35 (1981).

22. W.-T. Law and S. R. Crouch, *Anal. Lett.*, **13** (B3), 1115 (1980).

23. N. S. Pimenova, R. A. Ivanova, E. A. Zhiguleva, E. I. Kozlov, L. D. Ageeva, Yu. A. Koshelev, and V. F. Tsepalov, *Khim. Farm. Zh.*, **16**, 504 (1982); *Chem. Abstr.*, **96**, 205331n (1982).

24. S. U. Kreingol'd, L. V. Lavrelashvili, and I. M. Nelen', *Zh. Anal. Khim.*, **37**, 1853 (1982); *J. Anal. Chem. USSR (Engl. Transl.)*, **37**, 1441 (1982).

25. M. A. Koupparis, C. E. Efstathiou, and T. P. Hadjiioannou, *Anal. Chim. Acta*, **107**, 91 (1979).

26. S. U. Kreingol'd, V. N. Antonov, and E. M. Yutal, *Zh. Anal. Khim.*, **32**, 1618 (1977); *Chem. Abstr.*, **88**, 78493z (1978).

27. M. A. Ryan and J. D. Ingle, Jr., *Anal. Chem.*, **52**, 2177 (1980).

28. E. P. Diamandis and T. P. Hadjiioannou, *Clin. Chem. (Winston-Salem, N. C.)*, **27**, 455 (1981).

29. M. A. H. Elsayed, M. H. Barary, and H. Mahgoub, *Anal. Lett.*, **18** (B11), 1357 (1985).

30. A. G. Xenakis and M. I. Karayanis, *Anal. Chim. Acta*, **159**, 343 (1984).

31. Y. R. Tahbouh and H. L. Pardue, *Anal. Chim. Acta*, **173**, 23 (1985).

32. K. Bartl, J. Ziegenhorn, I. Streitberger, and G. Assmann, *Clin. Chem. (Winston-Salem, N. C.)*, **128**, 199 (1983).

33. S. U. Kreingol'd, L. I. Kefilyan, and V. N. Antonov, *Zh. Anal. Khim.*, **32**, 2424 (1977); *Chem. Abstr.*, **88**, 78493z (1978).

34. K. Hiromi, H. Fujimori, J. Yamaguchi-ito, H. Nakatani, M. Onishi, and B. Tonomura, *Chem. Lett.*, 1333 (1977).

35. A. Tabacco, F. Bardelli, F. Meiattini, and P. Tarli, *Clin. Chim. Acta*, **104**, 405 (1980).

36. H. Tanigaki, H. Obata, and T. Tokuyama, *Bull. Chem. Soc. Jpn.*, **53**, 3195 (1980).

37. M. A. Elsayed, *Pharmazie*, **34**, 569 (1979).

38. H. Steinhart, *Anal. Chem.*, **51**, 1012 (1979).

39. M. A. Koupparis, K. M. Walczak, and H. V. Malmstadt, *J. Pharm. Sci.*, **68**, 1479 (1979).

40. J. K. Baker, *Anal. Chem.*, **54**, 347 (1982).

# DIFFERENTIAL REACTION RATE METHODS

This chapter is devoted to the simultaneous determination of two or more chemical species by what has come to be known as differential reaction rate methods. They constitute one of the two main branches of kinetic-based determinations; catalytic methods constitute the other (1). As used here, the word *differential* has no mathematical connotation; it refers only to differentiating among chemical species via rate measurements without prior separation. Determinations based on processes other than chemical reactions (e.g., radioactive decay, mass transfer) are not included here.

The reaction rates of closely related components of a mixture reacting with a common reagent are often quite similar. These rates cannot be sufficiently separated by either a thermodynamic or a kinetic masking technique to allow the faster- or slower-reacting component to be neglected. Differential reaction rate methods many times offer an alternative to physicochemical separation in such cases.

In contrast to catalytically based methods, differential rate determinations are not aimed primarily at determining low concentrations of materials in solution, and most of the reactions used are uncatalyzed. Although most applications of differential reaction rate methods are for the determination of organic species, a review on the determination of inorganic species is available (2).

Although more useful than many reported catalytic determinations, reports of differential rate determinations are comparatively scarce in the literature. This reflects the fact that conditions for determining more than one species simultaneously are difficult to design and that indicator reactions for single-species determination (particularly of transition metal ions by redox reactions) provide an almost endless list of new systems.

Let us consider the irreversible bimolecular reaction of a binary mixture of components A and B with a common reagent R:

$$A + R \xrightarrow{\ k_A\ } P \tag{7.1}$$

$$B + R \xrightarrow{\ k_B\ } P' \tag{7.2}$$

The range of concentrations of reactant and reagent for which general differential methods have been developed is shown in Figure 7.1. Besides of the kinetic regions shown in the figure, there are differential reaction rate methods based on the measurement of initial rates in which the kinetics becomes pseudo-zero-order.

Theoretical treatments for analysis based on second-order kinetics are considerably more involved than those for first-order or pseudo-first-order. Hence, whenever possible the conditions for a bimolecular reaction are adjusted so that pseudo-first-order kinetics is followed. A 50-fold or greater excess of reagent or reactants, corresponding to regions I and VII respectively in Figure 7.1, provides such a condition. There are systems, however, for which pseudo-first-order conditions cannot be employed and regions II to VI might need to be considered.

Differential methods fall into two broad categories depending on the approach used: (a) those based on *graphical evaluations*, and (b) those based on *mathematical computations*. There are features common to each approach, however, and in some cases the classification must be based on which one predominates. Logarithmic extrapolation (graphical approach) and the method of proportional equations (mathematical computations) are the most commonly encountered in practical applications. The main computational approaches have been organized in tabular form in Ref. 1 and are considered in detail in what follows.

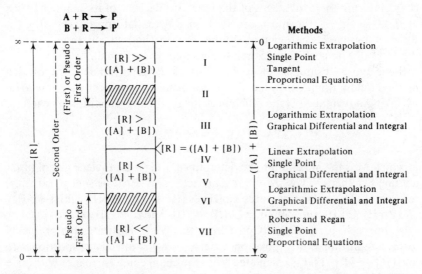

**Figure 7.1.** Classification of differential reaction rate methods based on reaction order. [From Ref. 45, p. 79. Reproduced with permission.]

## 7.1. METHODS BASED ON FIRST-ORDER OR PSEUDO-FIRST-ORDER KINETICS (REGION I OF FIGURE 7.1)

### 7.1.1. Logarithmic Extrapolation Method

Consider reactions 7.1 and 7.2. If $[R]_0 \gg [A]_0 + [B]_0$ and $P = P'$, the concentration of product P at any time $t$ is given by

$$[P]_\infty - [P]_t = [A]_t + [B]_t = [A]_0 e^{-k_A t} + [B]_0 e^{-k_B t} \qquad (7.3)$$

where $[P]_\infty$ is the concentration of product when reactions 7.1 and 7.2 have been completed. We assume A to react fast ($k_A \gg k_B$); thus, after A has essentially disappeared ($[A]_t \to 0$), we can take the logarithm of both sides of eq. 7.3 and obtain

$$\ln([P]_\infty - [P]_t) = \ln([A]_t + [B]_t) = -k_B t + \ln[B]_0 \qquad (7.4)$$

A plot of $\ln([A]_t + [B]_t)$ or $\ln([P]_\infty - [P]_t)$ versus time yields a straight line with a slope of $-k_B$ and an intercept (at $t = 0$) of $\ln[B]_0$. The value of $[A]_0$ may then be obtained by subtracting $[B]_0$ from the total initial concentration of A and B in the mixture; the latter can be determined by independent methods or can be calculated from $[P]_\infty$, provided the reaction mechanism does not change during the final stages of the reaction. Because of its simplicity, this method is one of the most widely used differential approaches; it gives somewhat greater accuracy than do the other methods for mixtures in which the ratio of rate coefficients is relatively large or very small.

Some typical examples of applications of the logarithmic extrapolation method follow. Some heavy metal ions, for instance, have been determined (3) by taking advantage of rate differences for the ligand substitution reaction

$$M–EGTA + PAR \rightleftharpoons M–PAR + EGTA$$

in which M is the metal ion to be determined, EGTA is ethylene glycol bis(2-aminoethylether)-$N,N,N',N'$-tetraacetic acid, and PAR is 4-(2-pyridylazo) resorcinol. The binary mixtures Co(II)–Ni(II), Cu(II)–Pb(II), Ni(II)–Fe(III), and Zn(II)–Cd(II), and a mixture Co(II)–Ni(II)–Zn(II)–Pb(II) were resolved by this approach. Figure 7.2, adapted from this work, illustrates the graphical form of logarithmic extrapolation. Log($A_\infty - A_t$) is directly proportional to log($[P]_\infty - [P]_t$) and is calculated from the change in absorbance with time for the Ni–PAR complex, the slower-reacting complex in mixture with Co(II). Logarithmic extrapolation has been applied by Hiraki et al. (4) to analyze

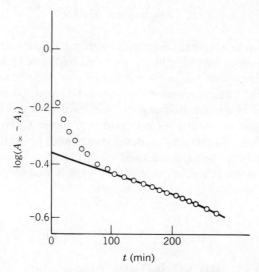

**Figure 7.2.** Plot of log $(A_\infty - A_t)$ as a function of time according to eq. 7.4. Determination of cobalt and nickel in a mixture containing Co(II), Ni(II), Zn(II), and Pb(II), all at the $10^{-6} M$ concentration level. Straight line is the calculated theoretical line. [Adapted from Ref. 3.]

binary mixtures of metal complexes (aluminum, gallium, indium, magnesium, zinc, and cadmium) of 5-sulfo-8-quinolinol in a method based on differences in fluorescence lifetimes. The method was applied to the determination of aluminum in magnesium alloy and magnesium in aluminum alloy. When the difference in fluorescence lifetime exceeds 4 ns, each component in the binary mixture can be determined from the logarithmic extrapolation plot resulting from analysis of the fluorescence decay curve. Onoue et al. (5) have applied the same approach to the simultaneous determination of 8-quinolinol and 5,7-dihalo-8-quinolinols. Logarithmic extrapolation can be used for the selective, simultaneous determination of carbenicillin and its pro-drugs carindacellin and carfecillin (6). The method is based on the different rates of degradation of carbenicillin and its carboxy ester derivatives in acidic solution (0.10 $M$ formate buffer, pH 3.00). The course of the reaction is followed by spectrophotometric monitoring using an imidazole determination. Mixtures of iron(III) and cobalt(II) can be resolved by exploiting the fact that complex formation of cobalt(II) with pyridoxal thiosemicarbazone is substantially faster than the corresponding reaction with iron(III) (7). Only simple mixing is required, making accessible the use of conventional spectrophotometric procedures for monitoring the course of the reaction. However, the single-point method (discussed in the next section) gave smaller errors and wider concentration ranges than the logarithmic extrapolation approach.

### 7.1.2. Single-Point Method

This approach can be applied to binary mixtures reacting irreversibly via first-order or pseudo-first-order kinetics and is based on the determination of the total concentration of analytes in the mixture by a single determination after a fixed reaction time. The approach was introduced by Kolthoff and Lee (8, 9) to resolve mixtures of unsaturated organic species in synthetic rubber using peroxybenzoic acid oxidation. As indicated in Chapter 1, this contribution played a vital role in encouraging the development and application of reaction rate measurements in chemical analysis.

If eq. 7.3 is divided by the sum of the initial concentrations of reactants being determined, we obtain

$$\frac{[A]_t+[B]_t}{[A]_0+[B]_0}=\frac{[P]_\infty-[P]_t}{[P]_\infty}=(e^{-k_At}-e^{-k_Bt})\frac{[A]_0}{[A]_0+[B]_0}+e^{-k_Bt} \quad (7.5)$$

Since $[A]_0/([A]_0+[B]_0)$ is the mole fraction of A, a plot of $([A]_t+[B]_t)/([A]_0+[B]_0)$ or $([P]_\infty-[P]_t)/[P]_\infty$ at any time versus the initial mole fraction of A in the mixture should give a straight line with a slope of $\exp(-k_At)-\exp(-k_Bt)$ and intercepts $\exp(-k_At)$ and $\exp(-k_Bt)$ at mole fractions of 1 and 0 respectively. The actual calibration curve (Figure 7.3) is easily constructed by measuring the extent of reaction of pure A and pure B at a given selected time $t$; a straight line is then drawn between the resulting two points. A convenient alternative is to calculate the intercepts from a knowledge of the corresponding rate coefficients for pure A and pure B. For each case

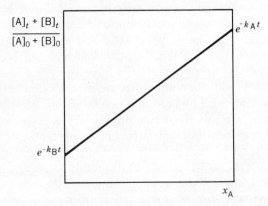

**Figure 7.3.** Single-point method for first-order reactions. $x_A$ = mole fraction of species A.

there is an optimum time for measurement that can be estimated (10) from

$$t_{opt} = \frac{\ln(k_A/k_B)}{k_A - k_B} \qquad (7.6)$$

Other criteria besides eq. 7.6, however, must be taken into consideration in some cases (11). Closely related pair of esters (e.g., ethyl acetate and isopropyl acetate) were determined with an error of 2% by use of their saponification rates (10). The single-point method, which can be considered to be a special case of the method of proportional equations (11), which is discussed later in this chapter, seems to provide better results when the ratio of rate coefficients for faster-reacting component over slower-reacting component is larger than 4.

Binary mixtures of phenols in water samples were analyzed by a modification of the single-point method discussed here (12). The modification consisted of taking into account the formation of two different final products in the oxidative coupling reaction with $N,N'$-diethyl-$p$-phenilenediamine caused by hexacyanoferrate(III). Stopped-flow mixing was used in these determinations.

### 7.1.3. Tangent Method

This method, also called the reaction rate method, is based on the determination of the rate of change of total concentration $[A]_t + [B]_t$ or $([P]_\infty - [P]_t)$ of the mixture (13). Differentiation of eq. 7.3 with respect to time yields

$$-\frac{d([P]_\infty - [P]_t)}{dt} = k_A[A]_t + k_B[B]_t \qquad (7.7)$$

If A is the faster-reacting component of the binary mixture, as its concentration approaches zero, the left-hand side of eq. 7.7 approaches a value equal to $k_B[B]_t$, and the rate of change of concentration can be determined at any time $t$ by plotting $[P]_\infty - [P]_t$ versus $t$ and estimating the slope at $t$ (Figure 7.4). As $[A]_t$ decreases, the ratio of the slope to $[P]_\infty - [P]_t$ approaches a constant value equal to $k_B$:

$$-\frac{d([P]_\infty - [P]_t)}{dt}\left(\frac{1}{[P]_\infty - [P]_t}\right)_{[A]_t \to 0} = k_B \qquad (7.8)$$

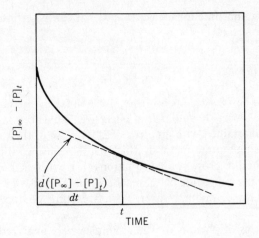

**Figure 7.4.** Determination by the rate of change of concentration.

A knowledge of $k_B$ permits calculation of $[B]_0$ from

$$\ln[B]_0 = \ln[B]_t + k_B t = \ln([P]_\infty - [P]_t) + k_B t \qquad (7.9)$$

The concentration of the faster-reacting component A must be determined from an independently measured value of $[A]_0 + [B]_0$ or from $[P]_\infty$. Application of this method has not been favored by practitioners; its real advantages surface when a continuous curve of total concentration change (e.g., $[P]_\infty - [P]_t$ vs. time) is available from an automatically recorded instrumental measurement.

### 7.1.4. Method of Proportional Equations

This approach, based on the principle of constant fractional life, has gained in popularity in most recent years. Two factors seem to account for this: (a) the use of metal (ligand)-exchange reactions, and (b) the use of digital computers for processing data. The high speed of most exchange reactions requires rapid mixing, which has been provided by using of the stopped-flow approach. The use of computers allows simultaneous determinations of $n$ species by employing a number of measurements $m$ such that $m \gg n$, a situation that results in better precision and accuracy (14).

Since the initial concentration of a species reacting with constant fractional life is directly proportional to the amount of product formed at any given time, for

$$A \xrightarrow{k_A} nP \qquad (7.10)$$

we have

$$[P]_t = n([A]_0 - [A]_t) = n[A]_0 (1 - e^{-k_A t}) = \phi[A]_0 \qquad (7.11)$$

where $\phi = $ constant.

Since any parameter $\psi$ that is directly proportional to concentration (e.g., absorbance, electrode potential, or diffusion current) can be measured,

$$\psi = a[P]$$

where $a = $ the proportionality constant between concentration and the measured physical property $\psi$, then

$$\psi_t = K_A[A]_0$$

where $K_A = a\phi$. For two species A and B reacting in the same fashion,

$$[P]_{t_1} = \phi_{A_1}[A]_0 + \phi_{B_1}[B]_0 \qquad (7.12)$$

$$[P]_{t_2} = \phi_{A_2}[A]_0 + \phi_{B_2}[B]_0 \qquad (7.13)$$

Numerical values for the proportionality constants $\phi$ need to be determined separately with pure A and pure B. Although knowledge of stoichiometric ratios, reaction rate coefficients, and the time at which measurements will be performed should permit calculation of the proportionality constants, the empirical approach is preferred to minimize the effect of experimental variables; what actually is determined are the individual values of $K$. Simultaneous solution of eqs. 7.12 and 7.13 yields the initial concentrations of A and B. The sum of the initial concentrations of the reactants does not need to be known; consequently the method can be applied to any number of chemical species reacting with a single reagent and to mixtures of species forming various reaction products with a common reagent. A variety of experimental variables have been used as discriminating variables to formulate simultaneous equations for application of the method of porportional equations. Typical examples are time (15–19), solvent(s) (20), temperature (21), concentration of enzyme catalyst (22), reaction conditions (23), and physicochemical characteristics of the components (24, 25). The fact that almost any experimental variable can be utilized to formulate the simultaneous equations is one of the attractions of the method of proportional equations. As an example of the many avenues open to application of this method, the use of a fluorescence quenching approach for the determination of two or three components in solution can be singled out (26). The method is based on the Stern–Volmer equation, which describes a dynamic quenching as follows:

$$F_0/F = 1 + K[Q]$$

where $F_0/F$ = ratio of fluorescence intensities in the absence and in the presence of quencher, $[Q]$ = quencher concentration, and $K$ = constant equal to the product of the decay time and the bimolecular rate coefficient for the quenching process. If several quenchers act simultaneously, the overall quenching process can be accounted for by adding additional terms to the Stern–Volmer equation:

$$F_0/F = 1 + K_1[Q_1] + K_2[Q_2] + \cdots + K_n[Q_n]$$

The method of proportional equations is implemented in this case by using $n$ independent equations obtained by measuring the ratio of fluorescence intensities in the presence of $n$ indicator species whose fluorescence is independently quenched by the quenchers to be determined. Applying this approach, Wolfbeis and Urbano (26) have simultaneously determined chloride and bromide in organic materials and chloride, bromide, and iodide in a synthetic mixture. Quinine, acridine, and harman in sulfuric acid media served as indicators since their fluorescence is dynamically quenched by the halides mentioned. A selection of analytical reaction rate methods developed using proportional equations is presented in Table 7.1.

Although theoretically there is no limitation on the number of components that could be determined by application of the method of proportional equations, a practical limit seems to be three at most four components. Errors in the determination of the proportionality constants and in the experimentally derived measurements impose the restriction. Examination of errors in analysis of systems with just two components (35) shows that the relative error for the determination of components A is given by

$$\frac{d[A]_0}{[A]_0} = \frac{d\psi(K_{B_1} + K_{B_2})}{\psi_2 K_{B_1} - \psi_1 K_{B_2}} \tag{7.14}$$

A similar expression can be derived for the relative concentration error for B. Equation 7.14 clearly indicates that the relative concentration errors depend on (a) the uncertainty in the measurement of the physical parameter, $d\psi$; (b) the absolute values of the individual measurements of the physical parameter; and (c) the values of the two proportionality constants.

**Table 7.1. Typical Applications of Method of Proportional Equations**

| Species Determined | Comments | Reference |
|---|---|---|
| Mixture of five organic peroxides (two peroxycarboxylic acids, two diacyl peroxides, and a hydroperoxide) | Use of a variety of reaction conditions which coupled with computer evaluation of five linear equations with five unknowns allowed determinations of as low as 0.002 meq of individual peroxide per milliliter. Relative errors <12%. | 25 |
| Ternary mixtures of Dy, Ho, and Yb | Stopped-flow spectrophotometry of ligand exchange between EDTA and the colored complexes of the rare earth metals with xylenol orange; time used as discriminating variable. | 19 |
| Binary and ternary mixtures of sulfonephthalein dyes (cresol red–cresol purple, cresol red–phenol red, cresol red–cresol purple–phenol red) | Determinations based on selective oxidations of the dyes by periodate ion (pH 7–10) catalyzed by Mn(II). Differentiation accomplished by altering one or more experimental conditions (e.g., catalyst concentration, pH, wavelength, time). | 24 |
| Ru and Os | Change in concentration of reactants in the indicator reaction [Ce(IV)–As(III)] provides rate differential. | 27 |
| Glucose and fructose | Reaction with ammonium molybdate in acid medium. | 17 |
| Binary mixtures of Th, U(VI), Pu(IV), and Np(IV). | Ligand exchange between the complexes with DCTA and Arsenazo III in 0.1 to 0.3 $M$ nitric acid. Absorbance measurement at 665 nm at two predetermined times. DTPA was the attacking ligand when neptunium was present. | 28, 29 |
| Ampicillin, pivampicillin, and bacampicillin | Reaction with an imidazole reagent to produce UV-absorbing penicillenic acid derivatives. Discriminating variable: time. Maximum relative uncertainty for individual components: 5%. | 30 |
| (6S)- and (6R)-mecillinam epimers | Difference in rates of reaction with a glycine reagent (pH 9.9, 35°C) and photometric monitoring of reaction product at two different times. | 31 |

**Table 7.1** (*continued*)

| Species Determined | Comments | Reference |
|---|---|---|
| Mixtures of betamethasone and its 17-valerate ester | Selective determination based on the oxidation of the steroid 21-hydroxy group with methanolic cupric acetate. Resulting aldehyde is subsequently condensed with 3-methylbenzothiazol-2-one hydrazone in alkaline medium forming a highly absorbing azine at 394 nm, which is monitored. | 32 |
| Phosphate and silicate | Time used as discriminating variable. Stopped-flow mixing and photometric monitoring of rate of formation of the 12-heteropolymolybdates. | 33 |
| Iron and manganese | Hydrogen peroxide oxidation of 2-hydroxybenzaldehyde thiosemicarbazone catalyzed with different rate coefficients by Mn(II) and Fe(III). Determination of both metal ions in aluminium and copper alloys, beer, cheeses, and soils. | 34 |

## 7.2. METHODS BASED ON SECOND-ORDER KINETICS (REGIONS III, IV, AND V OF FIGURE 7.1)

### 7.2.1. Second-Order Logarithmic Extrapolation

Extension of the logarithmic extrapolation method to include reactions run under second-order conditions was introduced by Siggia and Hanna (36). Their method is based on the same principle of plotting an expression for concentration versus time that becomes linear when the faster-reacting component has reacted virtually to completion. However, the approach is more involved than for first- or pseudo-first-order kinetics.

Consider again the reactions illustrated by eqs. 7.1 and 7.2. We assume A is again the faster-reacting component and define $x$ as the amount of A, B, or R reacted in a given time $t$. Then, when all of A has reacted, $x_A = [A]_0$ and the following applies (37):

$$\ln\left(\frac{[R]_0 - [A]_0 - x_B}{[B]_0 - x_B}\right) = k_B t\{[R]_0 - ([A]_0 + [B]_0)\} + K \qquad (7.15)$$

where

$$K = \ln\left(\frac{[R]_0 - [A]_0}{[B]_0}\right)$$

The initial concentration of A is determined from plotting $\ln[([R]_0 - x)/([M]_0 - x)]$, where $[M]_0 = [A]_0 + [B]_0$ and $x = x_A + x_{B'}$ versus time, and from extrapolating the linear part of this dependence corresponding to the reaction of the slower-reacting component B. Then the value at $t = 0$ is

$$K = \ln\left(\frac{[R]_0 - [A]_0}{[M]_0 - [A]_0}\right)$$

Note that the initial total concentration of reactants, $[M]_0$, must be known from an independent measurement since it is needed to construct the plot (Figure 7.5). Siggia and Hanna applied this approach to resolve mixtures of alcohols by following the rate of reaction with acetic anhydride (36) as well as mixtures of several other organic species containing the same functional group (38–40). A modification of the Siggia and Hanna approach has been developed by Bond et al. (41) that permits resolving mixtures for which the total concentration of A and B cannot be determined by functional group analysis and A and B react with R with a different stoichiometry. Siggia and Hanna applied the approach to analyze mixtures of cyclotrimethylenenitramine and

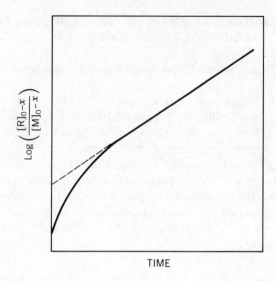

**Figure 7.5.** Second-order logarithmic extrapolation method.

cyclotetramethylenenitramine using their rate of hydrolysis in aqueous alkali. Papa et al. (42) have proposed a purely computational approach to this method that is somewhat less accurate than the graphical solution.

### 7.2.2.  Second-Order Linear Extrapolation (Region IV), Second-Order Single-Point (Region IV), and Second-Order Graphical Differential (Tangents) and Integral (Regions III, IV, and V) Methods

The second-order linear extrapolation method is applicable to irreversible second-order reactions (43). The single-point method applicable to irreversible second-order competitive reactions (10) is restricted to the very special case in which $[R]_0 = [A]_0 + [B]_0$ and requires knowledge of individual rate coefficients. The graphical differential and integral methods proposed by Schmalz and Geisler (44) are tedious, complex, and consequently of limited or no practical application beyond the original work of the proponents.

These general limitations are responsible for the fact that these methods have received little attention from a practical viewpoint. The reader interested in details on them is referred to the original papers or to two previous monographs on the general topic covered by this book (37, 45).

## 7.3. PSEUDO-FIRST-ORDER METHODS WITH RESPECT TO A COMMON REAGENT (REGION VII OF FIGURE 7.1)

### 7.3.1. Single-Point Method of Roberts and Regan

This method can be used with reactions for which it is possible to detect and follow changes in the concentration of R with high sensitivity. The rate of disappearance of R for the general case of eqs. 7.1 and 7.2 can be written

$$-\frac{d[R]}{dt} = k_A[A]_0[R] + k_B[B]_0[R] \tag{7.16}$$

Since $[R]_0 \ll [A]_0 + [B]_0$ and for all practical purposes $[A]_0$ and $[B]_0$ are constant, eq. 7.16 simplifies to

$$-\frac{d[R]}{dt} = k'_A[R] + k'_B[R] = K^*[R]$$

where

$$k'_A = k_A[A]_0, \; k'_B = k_B[B]_0$$

and

$$K^* = k_A[A]_0 + k_B[B]_0 \tag{7.17}$$

Moreover,

$$[M]_0 = [A]_0 + [B]_0 \tag{7.18}$$

Determinations are accomplished by solving eqs. 7.17 and 7.18 simultaneously; this requires an independent measurement of $[M]_0$ and determinations of $k_A$ and $k_B$ from runs with pure A and B. The value of $K^*$ is determined as usual for pseudo-first-order systems since

$$K^* = \frac{\ln([R]_0/[R]_1)}{t_1 - t_0}$$

The original method of Roberts and Regan (46) has been modified so that a single measurement and a calibration curve suffices for determination (47). Integration of eq. 7.16 gives

$$\ln\left(\frac{[R]_t}{[R]_0}\right) = -(k_A[A]_0 + k_B[B]_0)t = -K^*t$$

On substituting for $[B]_0$ and rearranging, this yields

$$\frac{[A]_0}{[M]_0} = \frac{\ln([R]_t/[R]_0)}{t[M]_0(k_B - k_A)} + \frac{k_B}{k_B - k_A} \tag{7.19}$$

Observation of eq. 7.19 permits the conclusion that at a given fraction, $[R]_t/[R]_0$, a plot of $1/t[M]_0$ versus the mole fraction of A, $([A]_0/[M]_0)$, yields the linear calibration curve. Minimal error in the determination requires estimation of the optimum fraction (optimum time interval), which is approximated (11) by

$$([R]_t/[R]_0)_{opt} = 1/e$$

This single-point version of the method of Roberts and Regan has been applied to the determination of mixtures of carbonyl compounds (47), hydroxy compounds (48), and amines (49).

### 7.3.2.  Method of Proportional Equations

The method of proportional equations has been modified to satisfy the condition where $[R]_0 \ll [A]_0 + [B]_0$ (50). The approach tolerates small ratios in rate coefficients (e.g., 2:1) and has been applied to the determination of several carbonyl compounds on reaction with hydroxylamine hydrochloride (50). The approach has several limitations (45). As the ratio of rate coefficients increases, the accuracy of the determination of the slower-reacting component deteriorates, and only mixtures that contain large amounts of such a component (or smaller amounts of the faster-reacting component) can be analyzed successfully.

### 7.4.  PSEUDO-ZERO-ORDER METHODS (REGIONS I THROUGH VII OF FIGURE 7.1)

These methods involve initial rate measurements for which it can be assumed that the concentrations of all species involved remain virtually constant during the measurement time and any value of their ratio is satisfactory. Advantages associated with initial rate methods (e.g., short time for determination and minimization of interference by side reactions) apply in these cases.

### 7.4.1.  Single-Point Method

The initial rate IR for the scheme represented in eqs. 7.1 and 7.2 can be represented by

$$IR = k_A^*[A]_0 + k_B^*[B]_0 \qquad (7.20)$$

where $k_A^* = k_A[R]_0$. Equation 7.20 and a prior knowledge of the total initial concentration of reacting species permits determination of $[A]_0$ and $[B]_0$ in the mixture. Binary mixtures of sugars have been analyzed by this approach (17).

### 7.4.2. Method of Proportional Equations

Except for involving initial rate measurements, the approach is entirely analogous to the method of proportional equations treated earlier in this chapter. Mark (22) applied it to the determination of alcohols in mixtures using enzyme-catalyzed reactions.

## 7.5. CRITICAL EVALUATION OF SELECTED DIFFERENTIAL REACTION RATE METHODS

This evaluation covers the most commonly used differential reaction rate methods and is based in part on Chapter 2 in Ellis's thesis (51). A comparison is summarized in Table 7.2. Examination of this table shows that the method of proportional equations, when applicable, is the best choice available to the analyst.

## 7.6. APPLICATION OF CONTINUOUS-FLOW SAMPLE–REAGENT(S) PROCESSING IN IMPLEMENTING DIFFERENTIAL REACTION RATE PROCEDURES

Differential reaction rate procedures have benefited by the flexibility for sample–reagent(s) processing that unsegmented-flow systems (52, 53) allow. A differential determination of magnesium and strontium by Dahl et al. (54) seems to be the first implementation of differential rate methods by continuous handling of sample and reagents. They used the well-known acid dissociation of *trans*-1,2-diaminocyclohexanetetraacetate metal complexes in the presence of Cu(II) ions as scavengers for the free ligand anion. The instrumental approach used was rudimentary but served to prove the point. It consisted of two spectrophotometers (Beckman DU) in tandem that monitored absorbance at 320 nm at two different times after initiating the reaction. The same approach was also used to determine magnesium and calcium with [2.2.1]cryptand as ligand and sodium ion as scavenger (55). Later on Kagenow and Jensen (56) proposed the simultaneous determination of strontium and magnesium (or calcium) ions by use of [2.2.2]cryptand this time; the manifold used consisted of only one detector in contrast with the dual detector approach of earlier contributions. A short review based on these works has

**Table 7.2. An Evaluation of Some Selected Differential Rate Methods**

| Advantages | Disadvantages |
|---|---|

*A. Graphical Extrapolation Methods (First Differential Rate Methods to Be Widely Used; Most Frequently Mentioned in the Earlier Literature)*

| Advantages | Disadvantages |
|---|---|
| 1. Since these methods depend on plotting the logarithm of the total reactant concentration against time, there is no need to determine rate coefficients. | 1. The faster-reacting component must be about 99% consumed before useful data can be obtained. |
| 2. Temperature is not a critical variable | 2. The total initial concentration of reactants must be known. Its determination often requires following the reaction to completion. |
| 3. The plotting procedure usually minimizes small errors since the "best" straight line is drawn through several points. | 3. If continuous monitoring of the reaction is not possible, a rather large number of samples has to be withdrawn from the mixture to follow the progress of the reaction. |
| 4. Since the methods are not restricted to constant processes following first-order kinetics, they can be applied in some cases in which synergism is observed. | |
| 5. The first-order method can be used for the determination of three components in a mixture. | |

*B. Method of Proportional Equation (Most Flexible Approach to Reaction Rate Determination. Though Its Application Is Rather Recent, It Is Now More Frequently Used.)*

| Advantages | Disadvantages |
|---|---|
| 1. Requires generally shorter times than the other approaches. | 1. The monitored property must be additive; hence the method is not applicable in case of synergism. |
| 2. Even if the mixture reacts by complex kinetics, it can be applied if the proportionality constants can be determined. | 2. Rate coefficients must be carefully measured. |
| 3. Prior knowledge of the total initial concentration of reactans is not required. | |
| 4. Values of ratios of rate coefficients $\simeq 4$ suffice. | |
| 5. Easily adaptable to automation, being ideal for fast reactions and for initial rate measurements. | |

6. Readily adaptable to the determination of more than two components in a mixture. Preferably used in combination with a minicomputer since the formulation and solution of the necessary simultaneous equations is then a simple task.

## C. Method of Roberts and Regan

| | |
|---|---|
| 1. Small (or large) ratios of rate coefficients can be tolerated. In a mixture containing 2.5% of the fastest-reacting component ($k_A/k_B = 2.2$), for instance, A was determined with only 2 to 3% error. | 1. The method is restricted to binary mixtures. |
| 2. Since $[R]_0 \ll [A]_0 + [B]_0$, side reactions are minimized. | 2. The sum of initial concentrations of the reactants must be independently determined. |
| 3. Useful over a wide range of $[A]_0/[B]_0$ as long as $k_A[A]_0$ is about 5% of the value of $k_A[A]_0 = k_B[B]_0$. | 3. Useful only when it is possible to follow the concentration of reagent R or of a common product of the two competitive reactions. |
| 4. The large concentration of reactants can increase the rate of very slow reactions and make them useful for determination. This, of course, may also make some reactions too fast to be of use. | |

## D. First-Order Single-Point Method (Like Method of Proportional Equations, Is Based on the Constant Fractional Life Concept)

| | |
|---|---|
| 1. Applicable to nearly any reaction mechanism by using empirical calibration curves. | 1. Requires ratios of rate coefficients of 4:1 or larger. |
| | 2. The total initial concentration of reactants must be determined independently. |
| | 3. Limited to binary determinations. |

Source: H. A. Mottola, CRC Crit. Rev. Anal. Chem., 4, 229 (1975). Reproduced with permission. Copyright CRC Press, Inc., Boca Raton, FL.

been published (57), and in a subsequent paper the authors used a continuous-flow–stopped-flow approach (58).

A substantial contribution, which clearly illustrates the flexibility and simplicity of continuous-flow sample–reagents processing for differential reaction rate methods with a single detector, was advanced by Fernandez et al. (59). Three different manifolds are presented, discussed, and evaluated; they center on (a) splitting the sample and using parallel flow cells in a double-beam spectrophotometer, (b) splitting the sample and using a confluence point, and (c) using flow cells in series in a double-beam spectrophotometer. These configurations are illustrated in Figure 7.6. They have been utilized for the simultaneous determination of cobalt and nickel based on the different rates of complex formation with 2-hydroxybenzaldehyde thiosemicarbazone.

Lazaro et al. (60) have implemented sequential as well as differential catalytic-based fluorescence determinations for manganese and iron. These were determined in the ranges of 40 to about 500 ng/mL for manganese and 40 to 600 ng/mL for iron on the bases of the catalytic effect of the metals on the hydrogen peroxide oxidation of 2-hydroxybenzaldehyde thiosemicarbazone. Two approaches were tested; one used a diverting valve and selected the most convenient carrier for the determination of one ion in the presence of the other (sequential determination with two separate injections of sample). The second approach (simultaneous determination) used a configuration similar to manifold $A$ in Figure 7.6. Precision seems to be better in the sequential method, but sample acceptance rates are comparable (about 25 to 30 samples per hour) for both approaches.

Simultaneous determination of thiocyanate and iodide based on their catalysis of the Ce(IV)–As(III) indicator reaction and selective inactivation of $SCN^-$ by reaction with Ce(IV) in absence of As(III) has been conducted in a continuous-flow system (61). The method uses a double injection technique; different residence times for the two equal-sized and simultaneously injected samples, which are also subject to different chemical treatment within the flow manifold, yield two serial peaks. The first peak corresponds to iodide and the second peak to the mixture of thiocyanate and iodide.

The continuous-flow, closed-loop configuration proposed by Ríos et al. (62) offers interesting potential uses for differential rate determinations since it allows measurements on the same sample plug at different times (e.g., multiple detection with a single detector on line).

### 7.7. SOME MISCELLANEOUS APPROACHES TO DIFFERENTIAL DETERMINATIONS

A unique, time-resolved determination of sulfur compounds by emission spectroscopy has been described by Schubert et al. (63). Such determination in

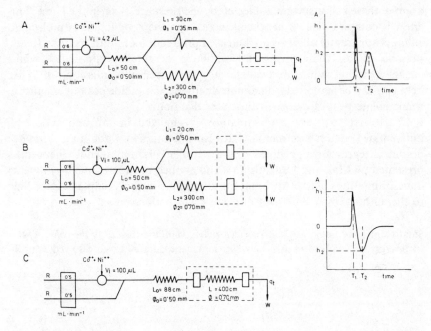

**Figure 7.6.** Continuous-flow manifolds with optimal physicochemical and sample transport variables (R is reagent in carrier): (*A*) Splitting up and confluence points and a single-beam spectrophotometer; (*B*) Splitting up of the sample and two flow cells in a double-beam spectrophotometer; (*C*) series configuration. Two typical peaks obtained are shown on the right. $V_i$ = volume of injected sample; $L_0, L_1, L_2$: reactors length; $\phi_0, \phi_1, \phi_2$: tube diameter. [Reprinted with permission from A. Fernandez et al., *Anal. Chem.*, **56**, 1146, Copyright 1984 by the American Chemical Society.]

complex solid samples such as atmospheric particulates has been restricted to analysis for total sulfur. This is not satisfactory since different forms of inorganic sulfur play different environmental roles. Individual determinations of the four forms of routine interest (elemental sulfur, sulfide, sulfite, and sulfate) commonly requires elaborate procedures that start with dissolution of the solid sample. The simultaneous determination proposed by Schubert et al. is based on the different rates of molecular sulfur emission in an $H_2/N_2$ flame. The time interval between directly introducing the solid sample into the flame and observing the molecular sulfur emission maximum (monitored at 383.6 nm) is as follows: sulfite 1 to 5 s; elemental sulfur, 14 to 20 s; sulfide 25 to 31 s; sulfate without air, 35 s; sulfate with air, 11 to 19 s. Determination is based on log–log plots of emission intensity and weight of sulfur in the sample. This time-resolved molecular emission determination seems to be insensitive to differences in the associated cations. However, the simultaneous determination of these sulfur species is affected by the presence of air in the flame. The effect is complex since a beneficial increase in the production of molecular sulfur is apparently offset by an increased production of sulfur oxide(s). In the presence of air the response for sulfide is completely eliminated, while elemental sulfur, sulfite, and sulfate show an increase in emission rate. This behavior can be exploited to eliminate an interfering sulfide peak or enhance a weak sulfate peak. Figure 7.7 illustrates this point.

Of interest to those contemplating or engaged in the application of differential reaction rate methods are considerations on adapting standard nonlinear regression algorithms for the treatment of simultaneous kinetic data presented by Mak and Langford (64) and applied to the treatment of kinetic data for mixtures of free $Al^{3+}$ and the aluminum–citrate complex. According to the authors' own statement this work was performed as

an attempt to demonstrate that *provided reliable initial estimates of the values of the parameters to be fitted are ascertainable*, the kinetic data of a well resolved 2- or 3-

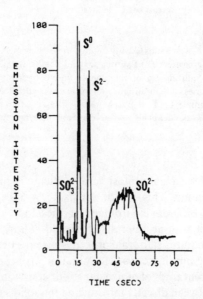

**Figure 7.7.** Characteristic emission responses of four inorganic forms of sulfur as a function of time. Relative sulfur concentrations due to sulfite ($SO_3^{2-}$), elemental sulfur ($S^0$), sulfide ($S^{2-}$), and sulfate ($SO_4^{2-}$) were approximately 1:1:8:9. Air was added to the flame mixture immediately after the sulfide peak was observed. Note the corresponding sharp dip about 28 to 29 s after sample introduction. [Reprinted from S. A. Schubert et al., *Anal. Chem.*, **52**, 963. Copyright 1980 by the American Chemical Society.]

component system where the respective rates observed are separated by nearly an order of magnitude can be treated satisfactorily by *any one* of the common non-linear regression methods.

Multipoint and nonlinear regression are treatments of data that with the accessibility of digital computers permit better utilization of experimental data and offer advantages of interest in method development.

Concurrent analysis of two- and three-component mixtures of amino acids (glycine–isoleucine, leucine–isoleucine, histidine–lysine–isoleucine, and histidine–leucine–isoleucine) has been accomplished by multipoint collection of rate information and nonlinear regression analysis of the data (65). A multipoint computer analysis of the rate profile was used in connection with a differential reaction rate approach to the determination of chlorine and chloramines in aqueous solution (66). The method is based on the reaction of the oxalate of $N,N$-diethyl-$p$-phenylenediamine with free chlorine, mono-chloroamines, and dichloroamines in the presence of iodide ion as catalyst in acidic solution. It utilized stopped-flow mixing; after mixing, the free chlorine reacts too fast for its contribution to the rate to be recorded, but its concentration can be estimated by the sudden change in absorbance at time "zero." In each run 256 data points were used.

Improved techniques to acquire, calculate, and test data sets for simultaneous kinetic analysis have been covered by Ridder and Margerum (67). In this treatment, emphasis is given to the importance of using the entire response curve with the aid of a minicomputer, instead of the graphical approach based on the signal after all but the slowest component has reacted, or initial rate measurements for the fastest component, or just two-point methods. Such considerations are illustrated with the simultaneous determination of the four-component mixture Zn(II)–Cd(II)–Hg(II)–Cu(II) at the $10^{-6}$ to $10^{-5}$ $M$ level. The procedure consisted in following the dissociation of metal–Zincon complexes at 620 nm (stopped-flow mixing). Each complex has a different molar absorptivity as well as a different rate coefficient for dissociation.

A graphical method that uses all the data to generate a linear plot giving the initial reactant concentrations has been proposed by Connors (68). Equation 7.3 can be rearranged as follows:

$$([P]_\infty - [P]_t)e^{k_A t} = [A]_0 + [B]_0 e^{(k_A - k_B)t} \tag{7.21}$$

Equation 7.21 shows that a plot of the left-hand side versus $\exp(k_A - k_B)t$ should be linear, with a slope of $[B]_0$. When $t = 0$, $\exp(k_A - k_B)t = 1$, and the ordinate has the value $[P]_\infty = [A]_0 + [B]_0$ and is referred as the intercept of the plot (Figure 7.8a). The slope/intercept ratio is equal to the original fraction of B in the mixture. Equation 7.21 can be expressed on the basis of fractions

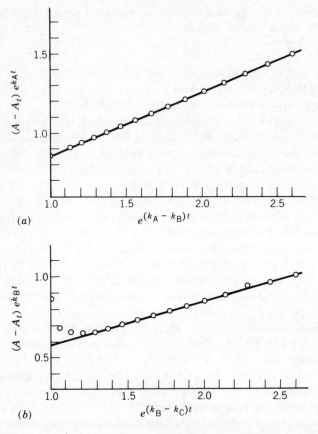

**Figure 7.8.** Linear graphical extrapolation (a) for a mixture of p-nitrophenyl p-chlorobenzoate and p-nitrophenyl benzoate, and (b) for a mixture of cyanobenzoate, chlorobenzoate, and benzoate esters. A = absorbance values, directly proportional to concentration. [Reprinted with permission from K. A. Connors, *Anal. Chem.*, **48**, 87. Copyright 1976 by the American Chemical Society.]

rather than concentrations by dividing each term by $[P]_\infty$. The main source of error in this graphical approach is in the $k_A$ and $k_B$ values needed to make the plot. They must be determined in separate experiments using pure components, and rate additivity must be assumed for the method to work in actual mixtures. The method was applied to mixtures of p-nitrophenyl esters of some *para*-substituted benzoic acids. At constant pH these esters undergo hydrolysis with pseudo-first-order kinetics. A three-component mixture (A, B, and C) could be handled by plotting $([P]_\infty - [P]_t)\exp(k_B t)$ versus $\exp(k_B - k_C)t$.

When all of A (the fastest-reacting component) has been used up, the plot becomes linear with a slope of $[C]_0$ and extrapolated intercept $[B]_0 + [C]_0$ (Figure 7.8$b$). The original concentration of A is derived from the knowledge that $[P]_\infty = [A]_0 + [B]_0 + [C]_0$.

If eq. 7.11 is written for the case of two species undergoing first-order reactions simultaneously as illustrated in eq. 7.1 and 7.2, and if the reactions are independent of one another, we have

$$[P]_t = [A]_0(1 - e^{-k_A t}) + [B]_0(1 - e^{-k_B t}) \qquad (7.22)$$

This is another form of eq. 7.3 and consequently eq. 7.21 written differently. Equation 7.22 indicates that if $k_A \gg k_B$, there will be an earlier time period during which $\exp(-k_B t)$ is close to unity; hence the quantity $1 - \exp(k_B t)$ is zero for all practical purposes, so that

$$[P]_t = [A]_0(1 - e^{-k_A t})$$

and plotting $[P]_t$ versus $\exp(-k_A t)$ permits one to determine the concentration of the faster-reacting component in the mixture. Similarly, later in the reaction profile the situation will be represented by

$$[P]_t = [A]_0 + [B]_0(1 - e^{-k_B t})$$

A plot of $[P]_t$ versus $\exp(-k_B t)$ will be a straight line with intercepts of $[A]_0$ and $[B]_0$ (at $t = 0$ and $t = \infty$ respectively), which permits calculation of $[B]_0$ by difference. This different way of examining and manipulating the basic equation 7.3 was used by Worthington and Pardue (69) in implementing a simple analog system providing automatic graphical presentation of the data. The instrumental aspects of this approach are considered in Chapter 9.

A graphical interpolation method (in contrast with all previously discussed methods, which mainly involve extrapolation) for the simultaneous determination of three components in a mixture has been proposed by Connors (70). The method applies for the general condition $[R]_0 \gg [A]_0 + [B]_0 + [C]_0$, requires no calculations or knowledge of the exact rate equations, and circumvents mutual kinetic effects. First determined is the concentration–time curve for the production of P for the *sample* mixture. Then synthetic *reference* mixtures are formulated and subjected to the same reaction conditions, again yielding concentration–time profiles for production of P in the reference mixtures. By successive adjustments in the initial reference concentrations of A, B, and C, a complete matching of the sample and reference concentration–time curves can finally be obtained, such that for all values of $t$, $[P]_t$ (sample) $= [P]_t$ (reference) and for such a reference mixture [A], [B], and

[C] are the same as in the sample solution. For a three-component mixture of p-nitrophenyl esters reacting with hydroxylamine, a graphical interpolation was applied (70) following the outlined rationale and using triangular coordinate diagrams to represent composition. Isokinetic contours were constructed on such diagrams to obtain values for the concentration of components in the sample. The triangular composition diagrams are logistically based on Roozeboom's triangular phase-behavior diagrams for three-component systems (71).

# REFERENCES

1. H. A. Mottola, *CRC Crit. Rev. Anal. Chem.*, **4**, 229 (1975).
2. D. Perez-Bendito, *Analyst (London)*, **109**, 891 (1984).
3. M. Tanaka, S. Funahashi, and K. Shirai, *Anal. Chim. Acta*, **39**, 437 (1967).
4. K. Hiraki, K. Morishige, and Y. Nishikawa, *Anal. Chim. Acta*, **97**, 121 (1978).
5. Y. Onoue, K. Morishige, K. Hiraki, and Y. Nishikawa, *Anal. Chim. Acta*, **106**, 67 (1979).
6. H. Bundgaard, *Arch. Pharm. Chemi., Sci. Ed.*, **7**, 95 (1979).
7. L. Ballesteros and D. Perez-Bendito, *Analyst (London)*, **108**, 443 (1983).
8. I. M. Kolthoff and T. S. Lee, *J. Polym. Sci.*, **2**, 200 (1947).
9. I. M. Kolthoff and T. S. Lee, *J. Polym. Sci.*, **2**, 220 (1947).
10. T. S. Lee and I. M. Kolthoff, *Ann. N. Y. Acad. Sci.*, **53**, 1093 (1951).
11. L. J. Papa, in *Kinetics in Analytical Chemistry*, H. B. Mark, Jr. and G. A. Rechnitz, Eds., Wiley, New York, 1968, Chapter 7.
12. E. Pelizzetti, G. Giraudi, and E. Mentasti, *Anal. Chim. Acta*, **94**, 479 (1977).
13. H. A. Laitinen and W. E. Harris, *Chemical Analysis*, 2nd ed., McGraw-Hill, New York, 1975, p. 392.
14. B. G. Willis, W. H. Woodruff, J. M. Frysinger, D. W. Margerum, and H. L. Pardue, *Anal. Chem.*, **42**, 1350 (1970).
15. R. G. Garmon and C. N. Reilley, *Anal. Chem.*, **34**, 600 (1962).
16. H. B. Mark, Jr., L. M. Backes, D. Pinkel, and L. J. Papa, *Talanta*, **12**, 27 (1965).
17. L. J. Papa, H. B. Mark, Jr., and C. N. Reilley, *Anal. Chem.*, **34**, 1443 (1962).
18. E. C. Toren, Jr. and M. K. Gnuse, *Anal. Lett.*, **1**, 295 (1968).
19. K. B. Yatsimirskii, A. G. Khachatryan, and L. I. Budarin, *Dokl. Akad. Nauk SSSR*, **211**, 1139 (1973).
20. H. B. Mark and R. A. Grienke, *J. Chem. Educ.*, **46**, 869 (1969).
21. D. Benson and N. Fletcher, *Talanta*, **13**, 1207 (1966).
22. H. B. Mark, Jr., *Anal. Chem.*, **36**, 1668 (1964).
23. J. D. Ingle and S. R. Crouch, *Anal. Chem.*, **43**, 71 (1971).

24. L. G. Ellis and H. A. Mottola, *Anal. Chem.*, **44**, 2037 (1972).

25. J. P. Hawk, E. L. McDaniel, T. D. Parish, and K. E. Simmons, *Anal. Chem.*, **44**, 1315 (1972).

26. O. S. Wolfbeis and E. Urbano, *Anal. Chem.*, **55**, 1904 (1983).

27. J. B. Worthington and H. L. Pardue, *Anal. Chem.*, **42**, 1157 (1970).

28. A. V. Stepanov, S. A. Nikitina, and T. A. Dem'yanova, *Radiokhimiya*, **21**, 34 (1979); *Chem. Abstr.* , **91**, 13085k (1979).

29. A. V. Stepanov, M. A. Nemtsova, S. A. Nikitina, and T. A. Dem'yanova, *Radiokhimiya*, **20**, 906 (1978); *Chem. Abstr.*, **90**, 161628j (1979).

30. H. Bundgaard, *Arch. Pharm. Chemi., Sci. Ed.*, **7**, 81 (1979).

31. H. Bundgaard, *Int. J. Pharmaceutics*, **5**, 257 (1980).

32. J. Hansen and H. Bundgaard, *Int. J. Pharm.*, **8**, 121 (1981).

33. C. C. Kircher and S. R. Crouch, *Anal. Chem.*, **55**, 248 (1983).

34. A. Moreno, M. Silva, and D. Perez-Bendito, *Anal. Chim. Acta*, **159**, 319 (1984).

35. K. Eckschlager, *Errors, Measurement and Results in Chemical Analysis*, Van Nostrand, London, 1969, p. 40.

36. S. Siggia and J. G. Hanna, *Anal. Chem.*, **33**, 896 (1961).

37. M. Kopanica and V. Stara, in *Wilson and Wilson's Comprehensive Analytical Chemistry*, Vol. 28, G. Svehla, Ed., Elsevier, Amsterdam, 1983, p. 139.

38. S. Siggia, J. G. Hanna, and N. M. Serencha, *Anal. Chem.*, **35**, 362 (1963).

39. S. Siggia, J. G. Hanna, and N. M. Serencha, *Anal. Chem.*, **35**, 365 (1963).

40. S. Siggia, J. G. Hanna, and N. M. Serencha, *Anal. Chem.*, **35**, 575 (1963).

41. B. D. Bond, H. J. Scullion, and C. P. Conduit, *Anal. Chem.*, **37**, 147 (1965).

42. L. J. Papa, H. B. Mark, Jr., and C. N. Reilley, *Anal. Chem.*, **34**, 1513 (1962).

43. C. N. Reilley and L. J. Papa, *Anal. Chem.*, **34**, 801 (1962).

44. E. O. Schmalz and G. Geisler, *Fresenius' Z. Anal. Chem.*, **188**, 241, 253 (1962); **190**, 222, 233 (1962).

45. H. B. Mark, Jr. and G. A. Rechnitz *Kinetics in Analytical Chemistry*, Wiley, New York, 1968, pp. 99–119.

46. J. D. Roberts and C. Regan, *Anal. Chem.*, **24**, 360 (1952).

47. L. J. Papa, J. H. Patterson, H. B. Mark, Jr., and C. N. Reilley, *Anal. Chem.*, **35**, 1889 (1963).

48. F. Willeboordse and R. L. Meeker, *Anal. Chem.*, **38**, 854 (1966).

49. R. A. Grienke and H. B. Mark, Jr., *Anal. Chem.*, **38**, 1001 (1966).

50. R. A. Grienke and H. B. Mark, Jr., *Anal. Chem.*, **38**, 340 (1966).

51. G. L. Ellis, Ph.D. thesis, "Kinetic and Equilibrium Simultaneous Analysis of Some Sulfonephthalein Dye Mixtures by the Method of Proportional Equations," Oklahoma State University, Stillwater, Okla., 1973.

52. J. Ruzicka and E. H. Hansen, *Flow Injection Analysis*, Wiley, New York, 1981.

53. M. Valcarcel and M. D. Luque de Castro, *Flow Injection Analysis: Principles and Applications*, Horwood, Chichester, U. K., 1987.

54. J. H. Dahl, D. Espersen, and A. Jensen, *Anal. Chim. Acta*, **105,** 327 (1979).

55. D. Espersen and A. Jensen, *Anal. Chim. Acta*, **108,** 241 (1979).

56. H. Kagenow and A. Jensen, *Anal. Chim. Acta*, **114,** 227 (1980).

57. D. Espersen, H. Kagenow, and A. Jensen, *Arch. Pharm. Chemi, Sci. Ed.*, **8,** 53 (1980).

58. H. Kagenow and A. Jensen, *Anal. Chim. Acta*, **145,** 125 (1983).

59. A. Fernandez, M. D. Luque de Castro, and M. Valcarcel, *Anal. Chem.*, **56,** 1146 (1984).

60. F. Lazaro, M. D. Luque de Castro, and M. Valcarcel, *Anal. Chim. Acta*, **169,** 141 (1985).

61. A. Tanaka, M. Miyazaki, and T. Deguchi, *Anal. Lett.*, **18** (*A*6), 695 (1985).

62. A. Ríos, M. D. Luque de Castro, and M. Valcarcel, *Anal. Chem.*, **57,** 1803 (1985).

63. S. A. Schubert, J. Wesley Clayton, Jr., and Q. Fernando, *Anal. Chem.*, **52,** 963 (1980).

64. M. K. S. Mak and C. H. Langford, *Inorg. Chim. Acta*, **70,** 237 (1983).

65. Y. Tahboub and H. L. Pardue, *Anal. Chim. Acta*, **173,** 43 (1985).

66. E. T. Gray, Jr. and H. J. Workman, in *Water Chlorination. Environmental Impact and Health Effect*, Vol. 4, Book 1, R. L. Jolley, W. A. Brungs, J. A. Cotruvo, R. B. Cumming, J. S. Mattice, and V. A. Jacobs, Eds., Ann Arbor Science, Ann Arbor, 1983, Chapter 48.

67. G. M. Ridder and D. W. Margerum, in *Essays on Analytical Chemistry (in Memory of Professor Anders Ringbom)*, E. Wanninen, Ed., Pergamon, Oxford, 1977, pp. 529–536.

68. K. A. Connors, *Anal. Chem.*, **48,** 87 (1976).

69. J. B. Worthington and H. L. Pardue, *Anal. Chem.*, **44,** 767 (1972).

70. K. A. Connors, *Anal. Chem.*, **49,** 1650 (1977).

71. A. Findlay, A. N. Campbell, and N. O. Smith, *The Phase Rule*, 9th ed., Dover, N.Y., 1951, p. 329.

# KINETIC METHODS BASED ON DETECTION OF LIGHT EMISSION: FLUORESCENCE, PHOSPHORESCENCE, CHEMILUMINESCENCE, AND BIOLUMINESCENCE

## 8.1. FLUORESCENCE AND PHOSPHORESCENCE

### 8.1.1. Fluorescence

Molecules of chemical species that absorb photons in the visible and ultraviolet regions of the spectrum are raised to excited electronic states. Most species dissipate this excess energy in the form of heat through collision with other species. Some species eliminate the excess by emitting light of a wavelength different from that of the absorbed photons, a process called *photoluminescence*. While photoluminescence includes Raman and Rayleigh scattering as well as fluorescence and phosphorescence, it is the latter two processes that are of most interest in kinetic-based analytical chemistry. Both are de-excitation processes of intrinsically kinetic nature. Their place in the excitation–de-excitation process is illustrated in Figure 8.1. Once a chemical species reaches the lowest vibrational level of an excited state, several things can occur. One is emission of a photon by return to the ground state, a process called *fluorescence*. On the other hand, radiation arising from triplet-to-singlet transition is termed *phosphorescence*.

During the de-excitation process a molecule can pass from a low vibrational level of the second excited singlet state $S_2$ to an equally energetic vibrational level of the first excited singlet state; the process is called *internal conversion* (IC). At the same time, while a molecule is in the excited state, one electron may reverse its spin. This corresponds to a multiplicity of 3: the molecule changes to a triplet state, and this process of nonradiative transfer is called *intersystem crossing* (ISC).

The kinetic nature of fluorescence and phosphorescence can be summarized as follows (1). The intensity of absorbed light, which can be equated to the rate of absorption $\Delta I$, for a fluorescence process can be expressed as

$$\Delta I = (k_{IC} + k_{ISC} + k_f + k_Q[Q])[S_1] \tag{8.1}$$

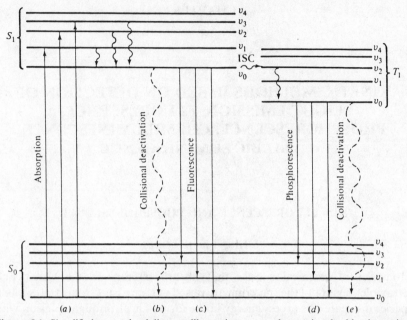

**Figure. 8.1.** Simplified energy-level diagram illustrating energy changes involved in absorption, fluorescence, and phosphorescence. $v$, vibrational levels; $S_0$, singlet ground state; $S_1$, singlet first excited state; $T_1$, triplet first excited state; ISC, intersystem crossing.

where $k_{IC}$, $k_{ISC}$, $k_f$, and $k_Q$ = rate coefficients for internal conversion/vibrational relaxation, intersystem crossing, fluorescence, and collisional quenching; [Q] = the quencher concentration; and $[S_1]$ = the steady-state concentration of molecules in the $S_1$ state. The fraction of the total number of photons absorbed that result in fluorescence can be defined as

$$\Phi_f = \frac{\text{total number of photons emitted}}{\text{total number of photons absorbed}} \tag{8.2}$$

This permits the following relation between rate of fluorescence $I_f$ and rate of absorption:

$$I_f = \Phi_f \Delta I = k_f[S_1] = \Phi_f(k_{IC} + k_{ISC} + k_f + k_Q[Q])[S_1] \tag{8.3}$$

Equation 8.3 leads to the following expression for the lifetime of the $S_1$ state, $\tau_f$:

$$\tau_f = \frac{1}{k_{IC} + k_{ISC} + k_f + k_Q[Q]} \tag{8.4}$$

A similar expression can be derived for phosphorescence:

$$\tau_p = \frac{1}{k_p + k'_{VR} + k_{Qp}[Q_p]} \tag{8.5}$$

where $k_p =$ the first-order rate constant for decay from $T_1$ to $S_0$, $k'_{VR} =$ the rate constant for the vibrational relaxation of the $T_1$ state, and $k_{Qp} =$ the rate constant for quenching of the triplet state by an impurity quencher $Q_p$. These considerations neglect the transferral process from the triplet to the singlet state leading to delayed fluorescence. Fluorescence or phosphorescence signals are measured under conditions of dynamic change controlled by the rates of de-excitation processes that follow the excitation.

Application of fluorescence to reaction rate methodology, however, is commonly considered only as a detection approach offering a competitive alternative, when it can be applied, to absorptiometric measurements in the UV–visual region and to electrochemical detection. The analytical advantages of this mode of detection in reaction rate methods are its high sensitivity (low limits of detection), selectivity, and wide dynamic range. For chemical species with a large value of $\Phi_f$, the limit of detection for fluorometric monitoring may be two to three orders of magnitude lower than for absorptiometric monitoring. Improvement by three to four orders of magnitude can also be realized in the range of concentrations amenable to determination. Changes in the fluorescence signal during the reaction are measured and related to the initial concentration of the analyte; the use of fluorometry as the monitoring approach in reaction-rate-based methods has been reviewed by Ingle and Ryan (2). These applications mostly concern catalyzed reactions (3). A typical example of such detection is the determination of iridium based on its catalytic effect on the classical Sandell–Kolthoff reaction with fluorescence monitoring of the Ce(III) produced in the indicator reaction (4). If the reaction does not involve a fluorescent reactant or product, determination can still be made possible by coupling with a fluorogenic species in a secondary indicator reaction, much as is commonly done in enzymatic determinations (5). For instance, the reduced form of nicotinamide adenine dinucleotide (NADH) is fluorescent while NAD is not. This property can be used to follow the progress of the main reaction by coupling enzyme-catalyzed reactions, as can be illustrated by fluorometric monitoring in glucose determination based on the following sequence of reactions (6):

$$\text{glucose} + \text{ATP} \xrightleftharpoons{\text{HK}} \text{ADP} + \text{D-glucose-6-PO}_4$$

$$\text{D-glucose-6-PO}_4 + \text{NAD} \xrightleftharpoons{\text{G-6-PDH}} \text{D-glucone-}\delta\text{-lactone-6-PO}_4 + \text{NADH}$$

where ATP = adenosine 5'-triphosphate, HK = hexokinase, ADP = adenosine 5'-diphosphate, and G-6-PDH = glucose-6-phosphate dehydrogenase. The NADH produced in the second reaction can react with nonfluorescent resazurin to produce a highly fluorescent resofurin and double the sensitivity (7).

The development of fluorescence methods has been reviewed by O'Haver (8). This development was greatly facilitated by advances in instrumentation in reaction rate methods (discussed in Chapter 9). The review by Ingle and Ryan (2) discusses instrumentation specifically for the use of fluorescence in rate measurements.

Both inorganic and organic species have been determined with fluorescence monitoring and by reaction rate procedures. A good example of determination of inorganic species is the determination of aluminum proposed by Wilson and Ingle (9). The determination is based on following the rate of formation of a fluorescent product in the reaction between aluminum(III) ions and 8-hydroxyquinoline-5-sulfonic acid. This illustrates the low limits of detection and large linear dynamic ranges afforded by fluorescence monitoring. The reported detection limit for Al(III) is 0.4 ppb; linear concentration plots can be obtained for up to 10 ppm with relative standard deviations typically less than 1%. The determination is quite free from interferences. In order to obtain results in an equilibrium-based determination one would need to wait about 30 min; thus the reaction rate procedure is faster and more selective.

A highly selective kinetic-fluorometric determination of vanadium at parts-per-million levels was proposed by Salinas et al. (10). Chemically the method relies on the oxidation of 1-amino-4-hydroxyanthraquinone by V(V), which produces a fluorescent product (excitation at 480 nm, emission at 575 nm). The initial rate approach (method of tangents) is recommended by the authors because it provides a broader dynamic range of applicability (100 to 530 ppm) and is more accurate than other methods. Only Ce(IV) exerts serious interference. The method was applied to the determination of nonvolatile vanadium in petroleums and found to compare well with atomic absorption and equilibrium-based colorimetric procedures.

The number of kinetic-based methods for determining alkaline earth ions is very limited because these ions do not participate in catalytic cycles. Laserna et al. (11) have proposed a selective determination of magnesium in microgram amounts based on the reaction, in basic medium, with 2-fluoroacetaldehyde 2-pyridylhydrazone (FAPH). At high pH values (e.g., 12.5 to 13.5) micellar magnesium hydroxide forms an intense green fluorescent species with FAPH. The fluorescence decreases with time and is monitored at 520 nm (excitation at 375 nm). A review of fluorometric reaction rate methods for the determination of inorganic species is available (12).

To fluoresce, an organic compound must contain highly conjugated bonds. Some vitamins offer such a structural characteristic, and measurement of the fluorescence of vitamins has found wide acceptance in the determination of such species. Once more a kinetic determination provides a much faster approach for determining, for instance, vitamin C by formation of its fluorescent adduct with o-phenylenediamine (2). The standard method requires over half an hour for equilibrium to be reached; a reaction rate determination requires only 16 s.

Steinhart (13) has developed a kinetic-fluorometric method for the determination of tryptophan based on its reaction with formaldehyde at a pH of 10.8 (carbonate buffer). The difference in fluorescence emission (excitation at 289 nm, emission at 356 nm) between 45.and 105 s, directly proportional to tryptophan concentration, was used for preparing calibration curves. Indole derivatives and guanine interfere, but since these species are rarely present in large amounts in food and feedstuffs, the method in most cases is suitable for determining tryptophan directly in such samples after alkaline hydrolysis. The limit of detection for tryptophan is reported as 2 nmol/mL.

### 8.1.2. Phosphorescence

As pointed out in the preceding section, phosphorescence shares its phenomenological origin and kinetic nature with fluorescence. The difference resides in the fact that phosphorescence offers a different path for the excited species to return to the ground state by emitting photons (see Figure 8.1). Four distinctive aspects distinguish phosphorescence from fluorescence (14):

1. Its decay periods are much longer (the decay from the triplet to the ground state singlet is forbidden by spin symmetry and is therefore slow).
2. Its spectrum is not the mirror image of the absorption spectrum.
3. It is not directly observed in solution at room temperature (the high collision rate between molecules deactivates the photoexcited state, and dissolved oxygen quenching can drastically limit phosphorescence in liquid state).
4. It is rarely observed in gases.

Points 1 and 3 are of particular impact on analytical methodology and are responsible for the comparatively limited application of phosphorescence until recently. It is commonly stated that phosphorescence has found more application in the study of the triplet state than as a determinative tool (14). However, this point needs qualification since in the past 15 years or so, room-temperature phosphorimetry has made inroads in chemical analysis (15). Thanks to the ingenuity of a handful of scientists, room-temperature phosphorimetry is no longer outside practical reach. It has found rich application in solid-phase as well as in liquid media (Table 8.1). As a matter of fact, room-temperature phosphorimetry is another example of an analytical approach in which adequate theoretical interpretation of the phenomena lags behind analytical applications.

The introduction of phosphorimetry into chemical analysis started with the work of Roth in 1967 (16). He observed strong room-temperature phosphorescence from a variety of organic species adsorbed on cellulosic supports. He attributed the emission to a surface effect. Later on Schulman and Walling (17, 18) investigated room-temperature phosphorescence on a variety of supports such as alumina, paper, and asbestos. They suggested that the rigidity of the solid support restricts collisional quenching and prevents radiationless deactivation of the triplet state. Other studies and applications have followed. The salient points for room-temperature phosphorimetry on solid supports may be summarized as follows:

1. Undoubtedly the adsorbed ("immobilized") state plays a critical role. No similar effect was observed from finely ground pure salts or from crystals grown from solution).
2. The adsorbed species (analyte) must be ionic in character (19).
3. Hydrogen bond interaction between the ionic analyte and polar groups in the support (e.g., –OH groups in cellulosic materials, alumina, and silica gel) plays a capital role (20). No phosphorescence was observed on the same surfaces after silylization.

**Table 8.1. Sample-Holding Media Employed in Room-Temperature Phosphorimetry**

*Solid Supports*

| | |
|---|---|
| Paper and chemically treated paper | Sodium acetate |
| Paper lint | Sucrose |
| Ion exchange paper | Starch |
| Alumina | Chalk |
| Silica gel | Other inorganic substances |
| Asbestos | Polymer–salt mixtures |

*Liquid Media*

Micellar systems
Sensitizer solutions

4. In the case of nonpolar species (e.g., polyaromatic hydrocarbons) the analyte–support bonding interaction involves only weak dispersion interactions (21), and the production of intense room-temperature phosphorescence emission from nonpolar aromatic compounds requires the use of heavy atoms in the medium (22).

Room-temperature phosphorescence in liquid media centers on emission in micelle-stabilized systems and in sensitized liquid systems (15). The first application of phosphorimetry in micelle-stabilized systems involved polynuclear aromatics in solution (23). Deoxygenation of the micellar system is required, as well as the use of heavy atoms in the form of silver or thallium decyl sulfate (24). In sensitized phosphorescence a weakly phosphorescent or nonphosphorescent species to be determined (donor) transfers its triplet energy to an added acceptor, producing an acceptor triplet with high phosphorescence yield. Donkerbroek et al. (25) reported the determination of benzophenone, 4-bromobenzophenone, 4-methylbenzophenone, 2-chlorobenzophenone, 4,4'-dibromobiphenyl, and 4-bromobiphenyl by sensitized room-temperature phosphorescence using biacetyl as acceptor and a 1 : 1 (v/v) mixture of acetonitrile and water.

A good example of the determinative capabilities afforded by the inherent kinetic nature of the phosphorescence process is by pulsed-source, time-resolved phosphorimetry. The technique was introduced by O'Haver and Winefordner in 1966 (26) and discussed in more detail by Winefordner three years latter (27). Using a laboratory-built pulsed-source phosphorimeter, Fisher and Winefordner (28) time-resolved several synthetic binary mixtures and a ternary mixture of phosphors with differing millisecond decays but otherwise very similar spectral characteristics. A modified version of the same unit was used by O'Donell et al. (29) to quantitate (by time resolution) binary and ternary mixtures of halobiphenyls; these species also had almost identical phosphorescence emission spectra but quite different phosphorescence lifetimes (roughly 5 to 500 ms), as shown in Figure 8.2. Instrumental modification involved the use of a rotating sample cell and a more intense excitation source. Using the rotating sample cell is expected to reduce the potential errors that may result from positioning the cell in the cell compartment. Figure 8.3 illustrates the components of the pulsed-source phosphorimeter (26).

Estimation of the concentration of analytes in mixtures is based on the differential rate of decay of the phosphorescence species in the mixture. As such the approach is a typical differential rate (although not differential reaction rate) kinetic method of determination. Thus the approaches proposed by the authors (26) for estimating concentrations are based on the method of proportional equations and on logarithmic extrapolation. The basic assumption of independent behavior (from an excitation–de-excitation point of view)

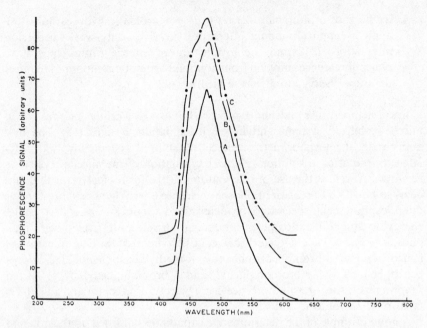

**Figure 8.2.** Phosphorescence spectra of some halogenated biphenyls. Solvent is ethyl alcohol. (*A*) 4-Chlorobiphenyl (lifetime: 570 ms). (*B*) 4-Bromobiphenyl (lifetime: 17 ms). (*C*) 4-Iodobiphenyl (lifetime: 3.2 ms). [Reprinted with permission from C. M. O'Donell et al., *Anal. Chem.*, **45**, 609, Copyright 1973 by the American Chemical Society.]

is necessary for quantitative application of the method of proportional equations, and it is desirable if logarithmic extrapolation is applied.

For a binary mixture of phosphors A and B with different phosphorescence lifetimes, the total phosphorescence measured at a (delay) time $t$ after the termination of excitation is given by

$$I_{tot} = I_{t_A} + I_{t_B} \tag{8.6}$$

The individual contributions are given by

$$I_{t_A} = I_{0_A} e^{-t/\tau_A} \tag{8.7}$$

and

$$I_{t_B} = I_{0_B} e^{-t/\tau_B} \tag{8.8}$$

where $\tau_A$ and $\tau_B$ are the measured phosphorescence decay times for component A and component B. For a system with multicomponent absorbers and

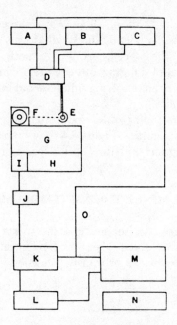

**Figure 8.3.** Block diagram of pulsed-source phosphorimeter. (*A*) Pulse generator. (*B*) 300-V power supply. (*C*) 1400-V power supply. (*D*) Trigger circuit. (*E*) Xenon flashtube compartment. (*F*) Sample cell compartment. (*G*) Emission monochromator. (*H*) Phototube power supply. (*I*) Photomultiplier. (*J*) Load resistors and fast switching diodes. (*K*) Readout (oscilloscope). (*L*) Preamplifier. (*M*) Signal averager. (*N*) Strip chart recorder. (*O*) Synchronization for oscilloscope and signal averager. [From Ref. 28. Copyright 1972 by the American Chemical Society, reproduced with permission]

phosphors, we can write

$$I_{tot} = \sum_i I_{t_i} e^{-t/\tau_i} \tag{8.9}$$

Equation 8.9 provides the basis for application of the methods of proportional equations and logarithmic extrapolation. A method of proportional equations can be applied, for instance, based on the linear-equation contributions of both components in the mixture:

$$I_{tot_1} = y_{A_1} + m_{A_1} C_A + y_{B_1} + m_{B_1} C_B$$
$$I_{tot_2} = y_{A_2} + m_{A_2} C_A + y_{B_2} + m_{B_2} C_B \tag{8.10}$$

where $I_{tot_1}$ and $I_{tot_2}$ are phosphorescence intensities of the mixture taken at different delay times $t_1$ and $t_2$, the $m$-values are the slopes and the $y$-values the corresponding intercepts. By solving the simultaneous equations in eq. 8.10, the values of $C_A$ and $C_B$, the concentrations of A and B in the mixture, can be estimated. Similarly, a set of simultaneous equations can be formulated based on eq. 8.9; the same approach of solving simultaneous equations should allow the estimation of $C_A$ and $C_B$.

Logarithmic extrapolation provides an avenue with fewer computational steps but somewhat less accuracy. Figure 8.4 shows typical logarithmic decay curves for binary and ternary mixtures of independently acting phosphors.

## 8.2. CHEMILUMINESCENCE AND BIOLUMINESCENCE

*Chemiluminescence* refers to the luminescence that arises during the course of a chemical reaction. It is observed when an electronically excited product is produced that either luminesces or transfers its energy to another molecule that then luminesces:

$$A + B \longrightarrow C^*$$
$$C^* \longrightarrow C + h\nu$$

or

$$C^* + D \longrightarrow D^* + C$$
$$D^* \longrightarrow D \quad \text{or} \quad D' + h\nu'$$

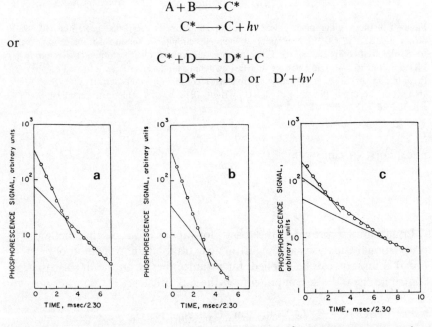

**Figure 8.4.** Logarithmic first-order decay curves for (*a*) $1 \times 10^{-3}$ $M$ anthrone and $1 \times 10^{-4}$ $M$ benzophenone; (*b*) $1 \times 10^{-3}$ $M$ anthrone and $1 \times 10^{-6}$ $M$ benzophenone; and (*c*) equimolar ($2.3 \times 10^{-6}$ $M$) mixture of 4-bromoacetophenone, benzaldehyde, and benzophenone. [From Ref. 28. Copyright 1972 by the American Chemical Society.]

In actuality, bioluminescence is a special case of chemiluminescence in which the chemical reaction responsible for photon emission occurs or can occur in nature. The *quantum yields* is the number (or rate) of molecules emitting light divided by the number (or rate) of molecules reacting. Typical quantum yields are about 1 to 5% in chemiluminescence but may approach 100% in very efficient bioluminescence reactions. A large number of chemi- and bioluminescence reactions are known, but few have found analytical application. Some recent reviews on chemi- and bioluminescence can be cited as sources for location of the most common, and a few not-so-common, reactions leading to light emission (30, 31).

Emission of light in the visible region of the spectrum requires free-energy changes of at least 44 kcal/mol, and this is usually the case for redox reactions (e.g., those involving oxidants such as oxygen or hydrogen peroxide and cyclic hydrazides such as luminol or acridine derivatives such as lucigenin) (31).

Although the number of reactions producing chemiluminescence is relatively small, metallic species increase the rate of chemiluminescence in some reactions, thus making the number of species amenable to determination not so limited. Rate-increasing effects have been reported for both organic and inorganic species (32). These effects are many times termed catalytic, but such designation is open to criticism since catalytic cycles are difficult to envision in most cases or the active species is consumed during reaction. Hence it would seem more appropriate to term these effects *promotion* (33).

The most frequently applied indicator reaction for metal ion determination based on promotion of chemiluminescence is the oxidation of luminol:

$$\text{luminol} + OH^- + \text{oxidant} \longrightarrow \text{product} + h\nu$$

luminol (5-amino-2,3-dihydrophthalazine-1,4-dione)

At least 20 metal ions increase the rate of chemiluminescence for this reaction; typical approximate limits of detection (in molar units) of $10^{-11}$ for Co(II), $10^{-9}$ for Cu(II), $10^{-8}$ for Ni(II), $10^{-9}$ for Cr(III), $10^{-10}$ for Fe(II), and $10^{-8}$ for Mn(II) have been reported (34). These low limits of detection make the system interesting for pollution studies, analysis of biological materials, and in metal ion detection in liquid chromatography. Haapakka and Kankare (35) explored the possibility of electrogenerating the $H_2O_2$; this eliminates the need for injection into the reaction zone or the use of flowing systems. Until recently only iron(II) was known to produce luminol chemiluminescence in the absence of $H_2O_2$. However, Klopf and Nieman (36) have reported that copper(II) and cobalt(II) also cause luminescence under the same conditions. On the other hand, manganese(II) produces very little, if any, such light emission in the absence of $H_2O_2$. Experimental conditions affecting luminol chemiluminescence in the presence and absence of $H_2O_2$ are also discussed by these authors. Metal ions are not the only species to alter the rate of oxidation of luminol by $H_2O_2$. Anionic species such as $OCl^-$ also act as promoters. In the absence of $H_2O_2$, however, $OCl^-$ also reacts with luminol, and a detailed study of interferences and optimization of the reaction has shown luminol to be suitable for the routine determination of $OCl^-$ in drinking water (37). Protein can also be determined with the luminol–$H_2O_2$ indicator reaction. However, Hara et al. (38) reported a 40-times greater sensitivity using the protein enhancement of the chemiluminescence generated in the 1,10-phenanthroline–$H_2O_2$ reaction. They developed a flow injection procedure able to handle 20 samples per hour with a minimum determinable quantity of 250 pg.

Luminol chemiluminescence can be sonically induced by propagation of ultrasonic waves through a basic aqueous solution of luminol containing dissolved oxygen. This has been used for the selective determination of as little as 0.07 pg of cobalt by its promoting effect (39). Ultrasonically induced luminescence provides a unique approach since it may require no reagents in the determination. Water in methanol, for instance, has been determined by ultrasonically induced chemiluminescence (40) but with lower sensitivity than with the Karl Fischer reagent.

As a rule, the detailed mechanisms of chemiluminescence reactions (including luminol oxidation) are not known or understood. This lack of knowledge is a handicap in analytical method development, control or even prediction of interferences is difficult. However, the factors determining the quantity and efficiency of light emission in chemiluminescence have been studied and reported (41). The lucigenin reaction has also been used for the determination of chemical species; limits of detection are considered to be several times lower with this reaction than with the luminol reaction (34). Montano and Ingle (42) have reported a determination of cobalt ions in solution with a very competitive limit of detection (20 ppt) based on the oxidation of lucigenin by $H_2O_2$ at pH 11 to 12. Linear calibration curves were obtained for up to 100 ppm of cobalt. The method was applied to the determination of cobalt in tap water (after dilution to eliminate interference from iron and magnesium) and in NBS standards for orchard leaves after digestion. The authors also developed a solvent extraction procedure to isolate cobalt from the most troublesome interfering species, although cobalt is the most effective promoter for the $H_2O_2$ oxidation of lucigenin (43). There are a few more potential indicator reactions, but further development work is needed to assess their analytical value.

Of significant analytical interest is the determination of gaseous species, such as nitrogen oxide, by gas-phase chemiluminescence based on the reaction

$$NO + O_3 \rightarrow NO_2 + O_2 + hv \qquad (\lambda_{max} = 600 \text{ nm})$$

Detection of NO can be accomplished at levels as low as 0.03 ppb (44). This reaction can also be used for the determination of nitrates, nitrites, and nitrosamines after reduction to nitrogen oxide (45–47). A comparison between the chemiluminescent determination of oxides of nitrogen and nitric acid and the determination of the same species by means of diode laser absorption techniques was conducted by Walega et al. (48). The study supports the preference for chemiluminescence measurements. Quite a few other species can be determined by gas-phase chemiluminescence. Nickel carbonyl, for instance, can be determined by the following reaction sequence (49):

$$Ni(CO_4) + O_3 \rightarrow NiO + products$$
$$NiO + CO \rightarrow Ni + CO_2$$
$$Ni + O_3 \rightarrow NiO^* + O_2$$
$$NiO^* \rightarrow NiO + hv \qquad (\lambda_{max} = 550\text{–}600 \text{ nm})$$

In keeping with the trend toward increasing use of immobilized reagents, some enzymes involved in bioluminescence reactions such as firefly luciferase, bacterial luciferase, and NAD(P)H: FMN oxidoreductase have been immobilized to glass (50, 51), Sepharose particles (52), or cellophane films (53).

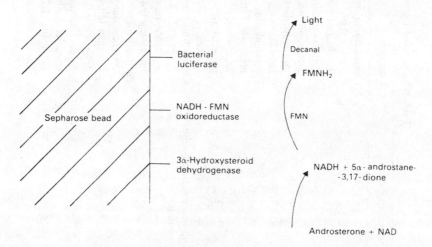

**Figure 8.5.** Reaction sequences in the determination of androsterone with three enzymes coimmobilized on Sepharose beads. [For details see Ref. 52.]

**Table 8.2. Selected Determinations Based on Chemi- or Bioluminescence**

| Determined Species | Comments | Reference |
|---|---|---|
| Copper | Chemiluminescence resulting from oxidative destruction (by $H_2O_2$ in basic medium) of 1,10-phenanthroline. The copper complex of 1,10-phenanthroline accelerates the process. Maximum emission at 445–450 nm. Cationic surfactant micelles increase sensitivity. Determination is highly selective. Detection limit is 15 pg/mL (competitive with nonflame atomic absorption and inductively coupled plasma emission spectrometry). Flow injection is used for reagent manipulation and transport to detection. | 54 |
| Cobalt | Electrochemical generation of luminescence from luminol at a rotating ring-disk electrode. Limit of detection is $10^{-5}$ $M$. Cr(III), Cu(II), and EDTA interfere. Discussion of mechanism of luminol electrogenerated chemiluminescence in aqueous alkaline solutions as produced with the rotating ring-disk electrode system. | 55/56 |
| $H_2S$ and other reduced-sulfur compounds | Chemiluminescence reaction with ozone. Chemiluminescence intensity decreases in the order $CH_3SH > CH_3SCH_3 > H_2S >$ thiophene. A commercially available, modified ozone monitor is used. The effect of oxides of nitrogen as increasing the signal is reported. | 57 |
| $H_2O_2$ | Peroxooxalate chemiluminescence. Use of fluorophore in immobilized form enhances detection. Controlled-pore glass is the most suitable solid support for immobilization. Implementation in a flow injection system. Limit of detection is 0.3 $\mu$g/mL. | 58 |
| $ClO_2(g)$ | Luminol chemiluminescence. A membrane flow cell is used to separate the reagents (luminol + $H_2O_2$, pH 10). Linear response is in concentration range of 0.050 to 0.20 $\mu$g/mL. Determination of $ClO_2$ in tap water, river water, and sewage effluent. Selective determination made in the presence of chlorine. Effects of phenol and chloramines under investigation. | 59 |

| Analyte | Description | Ref. |
|---|---|---|
| $SO_2(g)$ | Gas/solution chemiluminescence. Direct and continuous monitoring of $SO_2$ in atmospheres. Flavin mononucleotides enhance chemiluminescence of sulfite–permanganate reaction. Emission is further enhanced by polyoxyethylene(20), sorbitan trioleate micelle formation. Signal proportional to $[SO_2]^2$, detection limit is 3 $\mu$g/mL, and determinations are up to 30 $\mu$g/mL. | 60 |
| | $SO_2$ enhancement of the chemiluminescence reaction between $NO_2(g)$ and luminol in solution. 95% response of transient signal is obtained in 2 min. Limit of detection is 0.3 mg/mL. $H_2S$ and mercaptans strongly interfere. | 61 |
| $Br_2, Br^-$ | Luminol oxidation by $Br_2$. Determination made in water samples. | 62 |
| Glucose | Flow system in which an enzyme solution (glucose oxidase) flows, under pressure, through a microporous membrane allowing creation of a pH gradient in the flow cell. A 0.1 $M$ potassium phthalate buffer carries the enzyme across the membrane and assures a pH of 5 close to the membrane wall, which is optimum for the enzyme-catalyzed reaction. Sample is carried by a 0.5 $M$ KOH solution containing luminol and a metal complex "catalyst." pH in the bulk of the solution and close to the optical window for chemiluminescence measurement is about 11, optimal for the chemiluminescence reaction of luminol and $H_2O_2$. Simple addition of economical amounts of enzyme is the primary attraction of the approach. Application to the determination of glucose in serum after deproteinization and comparison with more conventional methods (nonchemiluminescence) and other chemiluminescence systems for glucose measurement reported. | 63 |
| Various metabolites | Species determined: D-glucose, L-lactate, 6-$P$-gluconate, L-malate, L-alanine, L-glutamate, NAD, NADP. Use of up to four coimmobilized enzymes on Sepharose 4B. Peak light intensity of bioluminescence of emitted light level at 460 nm is used as basis for determination. Wide range of concentrations are amenable to determination (pmol to nmol) range. | 64 |
| Oxalate and urate | Microamount determination using continuous-flow processing and column reactors packed with immobilized oxalate oxidase and uricase (two-column system) on porous glass beads. Chemiluminescence used to monitor $H_2O_2$ | 65 |

**Table 8.2.** (continued)

| Determined Species | Comments | Reference |
|---|---|---|
| | reacting with bis(2,4,5-trichlorophenyl) oxalate in presence of 9,10-diphenyl-anthracene and trimethylamine in a solution with freshly distilled dioxane as solvent. Determination range is $2 \times 10^{-5}$ to $8 \times 10^{-4}$ $M$. | |
| Uric acid, ascorbic acid | Measurement of pseudoinduction period. Iron-catalyzed oxidation of luminol by hydrogen peroxide. Use of corresponding oxidoreductase enzymes improves selectivity. Application is to aqueous standards and to blood serum and plasma. | 66 |
| $\alpha$-Amino acids | Flow injection system. Determination is based on inhibition of the Cu(II) accelerating effect on the luminol + $H_2O_2$ chemiluminescence. Minimum detection of L-aspartic acid is 2.7 ng per injection. Relative peak-area information given for 16 different amino acids. | 67 |
| Cholesterol | Cholesterol oxidase in solution (pH 7) forced through a microporous membrane into a flowing stream (2–10 mL/min) containing a plug of sample. The $H_2O_2$ produced reacts with luminol (pH 9, Tris buffer, and horseradish peroxidase as catalyst). Measurement of resulting chemiluminescence permits determination in range of 0.4–40 mg/dL. Detection limit is 0.2 mg/dL. Sample rate is 60 per hour, with only 0.01 units of enzymes used per determination. Application is to blood serum samples. | 68 |
| Bile acids | Use of 3-$\alpha$-hydroxysteroid dehydrogenase and a commercially available reagent containing low activity of NADH-FMN oxidoreductase and high activity of bacterial luciferase. Limit of detection is about 0.2 $\mu$mol/L. | 69 |
| | Specific determination of 12-$\alpha$-hydroxy bile acids. Use of enzymes coimmobilized on Sepharose 4B. Limit of detection is 4 pmol/0.5 mL of volume used in the measurement step. Linear range is 4–2000 pmol of bile acid. Good agreement with GLC determinations. Application is to determinations in serum of humans. | 70 |

164

| Analyte | Description | Reference |
|---|---|---|
| | Air-segmented continuous-flow system. A nylon reactor tube in which three enzymes have been coimmobilized was placed in front of a photomultiplier inside of luminometer. High selectivity for 7-$\alpha$-hydroxy bile acids; the overall reaction sequence is 7-$\alpha$-hydroxy bile acid $+ NAD^+$ $(E_1) \rightarrow$ 7-oxo bile acid $+ NADH + H^+$; $NADH + FMN + H^+$ $(E_2) \rightarrow NAD^+ + FMNH_2$; $FMNH_2$ $+ decanal + O_2$ $(E_3) \rightarrow$ decanoic acid $+ H_2O +$ light where $E_1 =$ 7-$\alpha$-hydroxysteroid dehydrogenase, $E_2 =$ NADH-FMN oxidoreductase, $E_3 =$ luciferase, FMN = flavin adenine mononucleotide. Linear response in the 10–2500 pmol range by measuring emitted light at 490 nm. Direct intercalation of serum samples permits processing 20–30 samples per hour. Nylon immobilized system superior to Sepharose-immobilized one. Reactor stable for more than 2 months (storage at 4°C between uses). | 71 |
| Heparin | Hydrolysis of heparin (4 h at 90°C in 0.25–0.50 $M$ $H_2SO_4$) to disaccharides, and there determined by chemiluminescence reaction with lucigenin. Minimum detection is 90 $\mu$g. Linear response of up to 10 mg of heparin. | 72 |
| Choriogonadotropin | Immunoassay for determination in human urine. Use of derivatives of isoluminol and measurement of chemiluminescence signal for 10 s. Labeled antigen is used with a second antibody covalently linked to polyacrylamide beads and excess of a specific antibody of different specificity. | 73 |
| Benzo[a]pyrene-7,8-dihydrodiol | Determination in complex metabolite mixture (reaction medium: $NaClO-H_2O_2$). Limit of detection is 30 pmol/mL. Results correlate well with those by high performance liquid chromatography (HPLC) determinations. | 74 |
| Adenosine 5'-triphosphate (ATP) | Flow injection system with specially built flow-through detector. Determination is based on use of firefly luciferin and luciferase. Useful response range for a 30-$\mu$L sample is $10^{-4}$ to $10^{-8}$ $M$ ATP. Sample rate is 200 samples per hour. | 75 |
| Reduced nicotinamide adenine dinucleotide [NADH] | Measurement of maximum bioluminescence intensity. Use of immobilized (on collagen strips) bacterial luciferase and flavin mononucleotide oxidoreductase. Immobilized system retains 70% of initial activity after 2 weeks. | 76 |
| Flavin adenine dinucleotide (FAD) | Determination based on the reactivation of apo-D-amino acid oxidase by FAD. In presence of D-alanine, luminol, horseradish peroxidase, and excess of the | 77 |

**Table 8.2.** (*continued*)

| Determined Species | Comments | Reference |
|---|---|---|
| | apoenzyme, FAD can be determined at the femtomole range (bicarbonate buffer, pH 9.2, and 37°C). | |
| Corticosteroids and *p*-nitrophenyl esters | Postcolumn chemiluminescence detection with lucigenin after HPLC separation. | 78 |
| Human and bovine serum albumin | Flow injection procedure based on the decrease of chemiluminescence (luminol + $H_2O_2$) by interaction between the proteinaceous material and the Fe(III) complex of $\alpha,\beta,\gamma,\delta$-tetraphenylporphinetrisulfonic acid. The iron complex accelerates chemiluminescence production. Optimum pH = 7.2. Minimum detectable quantities: 10 mg for human serum albumin and 4 mg for bovine serum albumin. | 79 |
| Diamines and polyamines | Hydrogen peroxide produced by oxidation of polyamines with serum aminase oxidase measured by luminol chemiluminescence. Determination is in the range of 10–100 pmol. Application is determination of urinary polyamines. | 80 |

166

Chemical modification of Sepharose 4B or CL6B by cyanogen bromide and subsequent immobilization of the enzyme has resulted in preparations with high activity. For instance, determination of picomolar levels of androsterone and testosterone have been possible with Sepharose chemically modified to coimmobilize 3-α-hydroxysteroid dehydrogenase, NADH-FMN oxidoreductase, and bacterial luciferase (52). The determination is based on the scheme illustrated in Figure 8.5. Table 8.2 lists recent and selected determinations based on chemi- and bioluminescence.

## REFERENCES

1. I. M. Warner, "Molecular Fluorescence and Phosphorescence," in *Instrumental Analysis*, 2nd ed., G. D. Christian and J. E. O'Reilly, Eds., Allyn and Bacon, Boston, 1986, Chapter 9.

2. J. D. Ingle, Jr. and M. A. Ryan, "Reaction Rate Methods in Fluorescence Analysis," in *Modern Fluorescence Spectroscopy*, E. L. Wehry, Ed., Plenum, New York, 1981, Vol. 3, Chapter 3.

3. H. A. Mottola and H. B. Mark, Jr., *Anal. Chem.*, **52**, 31R (1980); **54**, 62R (1982); **56**, 96R (1984); **58**, 264R (1966).

4. D. P. Shcherbov, O. D. Inyutina, and A. I. Ivankova, *Zh. Anal. Khim*, **28**, 1372 (1973); *J. Anal. Chem. USSR (Engl. Transl.)*, **28**, 1218 (1973).

5. H. A. Mottola, *Analyst (London)*, **112**, 719 (1987).

6. S. W. Kiang, J. W. Kuan, and G. G. Guilbault, *Clin. Chem. (Winston–Salem, N. C.)*, **21**, 1799 (1975).

7. G. G. Guilbault and D. N. Kramer, *Anal. Chem.*, **37**, 1219 (1965).

8. T. C. O'Haver, *J. Chem. Educ.*, **55**, 423 (1978).

9. R. L. Wilson and J. D. Ingle, Jr., *Anal. Chim. Acta*, **92**, 417 (1977).

10. F. Salinas, F. Garcia–Sanchez, F. Grases, and C. Genestar, *Anal. Lett.*, **13**, 473 (1980).

11. J. J. Laserna, A. Navas, and F. Garcia-Sanchez, *Microchem. J.*, **27**, 312 (1982).

12. M. Valcarcel and F. Grases, *Talanta*, **30**, 139 (1983).

13. H. Steinhart, *Anal. Chem.*, **51**, 1012 (1979).

14. R. Chang, *Basic Principles of Spectroscopy*, McGraw-Hill, New York, 1971, Chapter 12.

15. T. Vo-Dinh, *Room Temperature Phosphorimetry for Chemical Analysis*, Wiley-Interscience, New York, 1984.

16. M. Roth, *J. Chromatogr.*, **30**, 276 (1967).

17. E. M. Schulman and C. Walling, *Science*, **178**, 53 (1972).

18. E. M. Schulman and C. Walling, *J. Phys. Chem.*, **77**, 902 (1973).

19. R. T. Parker, R. S. Freedlander, and R. B. Dunlap, *Anal. Chim. Acta*, **119**, 189 (1980); **120**, 1 (1980).

20. E. M. Schulman and R. T. Parker, *J. Phys. Chem.*, **81**, 1932 (1977).

21. T. Vo-Dinh, E. Lue-Yen, and J. D. Winefordner, *Anal. Chem.*, **48**, 1186 (1976).

22. T. Vo-Dinh, E. Lue-Yen, and J. D. Winefordner, *Talanta*, **24**, 146 (1977).

23. L. J. Cline Love, M. Skrilec, and J. G. Habarta, *Anal. Chem.*, **52**, 754 (1980).

24. M. Skrilec and L. J. Cline Love, *Anal. Chem.*, **52**, 1559 (1980).

25. J. J. Donkerbroek, C. Gooijer, N. H. Velthorst, and R. W. Frei, *Anal. Chem.*, **54**, 891 (1982).

26. T. C. O'Haver and J. D. Winefordner, *Anal. Chem.*, **38**, 602 (1966).

27. J. D. Winefordner, *Acc. Chem. Res*, **2**, 361 (1969).

28. R. P. Fisher and J. D. Winefordner, *Anal. Chem.*, **44**, 948 (1972).

29. C. M. O'Donell, K. F. Harbaugh, R. P. Fisher, and J. D. Winefordner, *Anal. Chem.*, **45**, 609 (1973).

30. W. R. Seitz, *CRC Crit. Rev. Anal. Chem.*, **13**, 1 (1981).

31. L. J. Kricka and G. H. G. Thorpe, *Analyst (London)*, **108**, 1274 (1983).

32. U. Issacsson and G. Wettermark, *Anal. Chim. Acta*, **68**, 339 (1974).

33. V. V. S. Eswara Dutt and H. A. Mottola, *Anal. Chem.*, **46**, 1090 (1974).

34. W. R. Seitz and M. P. Neary, *Anal. Chem.*, **46** (2), 188A (1974).

35. K. E. Haapakka and J. J. Kankare, *Anal. Chim. Acta*, **118**, 333 (1980).

36. L. L. Klopf and T. A. Nieman, *Anal. Chem.*, **55**, 1080 (1983).

37. D. F. Marino and J. D. Ingle, Jr., *Anal. Chem.*, **53**, 455 (1981).

38. T. Hara, T. Ebuchi, A. Arai, and M. Imaki, *Bull. Chem. Soc. Jpn.*, **59**, 1833 (1986).

39. M. Yamada and S. Suzuki, *Chem. Lett.*, 783 (1983).

40. M. Yamada, T. Hobo, and S. Suzuki, *Chem. Lett.*, 283 (1983).

41. K. D. Gundermann, *Chemilumineszenz Organischer Verbindungen*, Springer-Verlag, Berlin, 1968.

42. L. A. Montano and J. D. Ingle, Jr., *Anal. Chem.*, **51**, 926 (1979).

43. L. A. Montano and J. D. Ingle, Jr., *Anal. Chem.*, **51**, 919 (1979).

44. B. A. Ridley and L. C. Howlett, *Rev. Sci. Instrum.*, **45**, 742 (1974).

45. R. C. Doerr, J. B. Fox, L. Lakritz, and W. Fiddler, *Anal. Chem.*, **53**, 381 (1981).

46. R. D. Cox, *Anal. Chem.*, **52**, 332 (1980).

47. K. S. Webb and T. A. Gough, *J. Chromatogr.*, **177**, 349 (1979).

48. J. G. Walega, D. H. Stedman, R. E. Shetter, G. I. Mackay, T. Iguchi, and H. I. Schiff, *Environ. Sci. Technol.*, **18**, 823 (1984).

49. D. H. Stedman, D. A. Tamaro, D. K. Branch, and R. Pearson, *Anal. Chem.*, **51**, 2340 (1979).

50. E. Jablonski and M. DeLuca, *Proc. Natl. Acad. Sci. USA*, **73**, 3848 (1976).

51. E. Jablonski and M. DeLuca, *Methods Enzymol.*, **57**, 202 (1978).

52. J. Ford and M. DeLuca, *Anal. Biochem.*, **110**, 43 (1981).

53. N. N. Ugarova, L. Yu. Brovko, and E. I. Beliaieva, *Enzyme Microb. Technol.*, **5**, 60 (1983).

54. M. Yamada and S. Suzuki, *Anal. Lett.*, **17**, 251 (1984).

55. K. E. Haapakka, *Anal. Chim. Acta*, **139**, 229 (1982).

56. K. E. Haapakka and J. J. Kankare, *Anal. Chim. Acta*, **138**, 263 (1982).

57. T. A. Lukovskaya, T. A. Bogoslovskaya, T. J. Kelly, J. S. Gaffney, M. F. Phillips, and R. L. Tanner, *Anal. Chem.*, **55**, 135 (1983).

58. G. Gubitz, P. van Zoonen, C. Goojier, N. H. Velthorst, and R. W. Frei, *Anal. Chem.*, **57**, 2071 (1985).

59. D. J. Saksa and R. B. Smart, *Environ. Sci. Technol.*, **19**, 450 (1985).

60. M. Kato, M. Yamada, and S. Suzuki, *Anal. Chem.*, **56**, 2529 (1984).

61. D. Zhang, Y. Maeda, and M. Munemori, *Anal. Chem.*, **57**, 2552 (1985).

62. A. T. Pilipenko, O. V. Zui, A. V. Terletskaya, *Zh. Anal. Khim.*, **38**, 1408 (1983); *J. Anal. Chem. USSR (Engl. Transl.)*, **38**, 1069 (1983).

63. D. Pilosof and T. A. Nieman, *Anal. Chem.*, **54**, 1698 (1982).

64. G. Wienhausen and M. DeLuca, *Anal. Biochem*, **127**, 380 (1982).

65. V. I. Rigin, *Zh. Anal. Khim.*, **37**, 1676 (1982); *J. Anal. Chem. USSR (Engl. Transl.)*, **37**, 1302 (1982).

66. J. E. Frew and P. Jones, *Anal. Lett.*, **18**, 1579 (1985).

67. T. Hara, M. Toriyama, T. Ebuchi, and M. Imaki, *Chem. Lett.*, 341 (1985).

68. N. Malavolti, D. Pilosof, and T. A. Nieman, *Anal. Chim. Acta*, **170**, 199 (1985).

69. I. Styrelius, A. Theore, and I. Bjorkhem, *Clin. Chem. (Winston–Salem, N.C.)*, **29**, 1123 (1983).

70. J. Schoelmerich, J. E. Hinkley, I. A. Macdonald, A. F. Hoffman, and M. DeLuca, *Anal. Biochem.*, **133**, 244 (1983).

71. A. Roda, S. Girotti, S. Ghini, B. Grigolo, G. Carres, and R. Bovara, *Clin, Chem. (Winston–Salem, N. C.)*, **30**, 206 (1984).

72. R. A. Steen and T. A. Nieman, *Anal. Chim. Acta*, **155**, 123 (1983).

73. G. J. Barnard, J. B. Klim, J. L. Brokelbank, W. P. Collins, B. Gaier, and F. Kohen, *Clin. Chem. (Winston–Salem, N. C.)*, **30**, 538 (1984).

74. A. Thompson, H. H. Seliger, and G. H. Posner, *Anal. Biochem.* **130**, 498 (1983).

75. P. J. Worsfold and A. Nabi, *Anal. Chim. Acta*, **171**, 333 (1985).

76. L. J. Blum and P. R. Coulet, *Anal. Chim. Acta*, **161**, 355 (1984).

77. A. Hinkkanen and K. Decker, *Anal. Biochem.*, **132**, 202 (1983).

78. M. Maeda and A. Tsuji, *J. Chromatogr.*, **352**, 213 (1986).

79. T. Hara, M. Toriyama, K. Kitamura, and M. Imaki, *Bull. Chem. Soc. Jpn.*, **58**, 2135 (1985).

80. U. Bachrach and Y. M. Plesser, *Anal. Biochem.*, **152**, 423 (1986).

# CHAPTER

9

# INSTRUMENTATION

Measurements of chemical systems evolving to equilibrium for analytical purposes require several modular units to perform specific tasks. These modular units are conceptually for (a) mixing of reactants, (b) monitoring of reaction progress (detection systems), (c) auxiliary electronics for signal conditioning and modification, and (d) computation and data handling. Sample and reagents must be mixed completely and rapidly and at constant temperature. The detection also must be made at constant temperature.

Each step indicated above is amenable to automation and thus kinetic measurements fit in the three-level general scheme of automation in the analytical laboratory (1). These three levels are:

*Level I*:   The sample as it arrives at the laboratory is logged in (coded) and is directly transferred to the second level (if appropriate) or prior to this is converted to a form compatible with the manipulations needed at the second level.

*Level II*:   The sample is introduced in a transporting system (e.g., robot-controlled conveyor or carrier stream in continuous-flow sample processing). The sample is then processed on-line as required (e.g., mixing with reagents mixed in, separations and/or preconcentrations performed). Detection is effected with the system at or approaching equilibrium by physical or chemical means, or both.

*Level III*:   The information contained in the analytical signal is modified and manipulated by the appropriate conversions and algorithms to obtain the final numerical results conveying the information sought.

The bulk of the operations performed in kinetic-based determinations are contained in levels II and III.

The mixing of reactants depends on the rate of change of the process being monitored and ranges from simple manual mixing to processing systems such as continuous-flow mixing. Monitoring chemical (sometimes physical) changes in kinetic determinations has been conducted mainly by absorptiometric methods (absorption of radiant energy mainly in the visible, occasionally in the UV, regions of the spectrum) and electroanalytical techniques. Implementation of level III is dominated by electronics and computers and makes heavy use of analog-to-digital and digital-to-analog conversion in the electrical domain and final transformation to numerical (or graphical) form in the nonelectrical domain (2, 3).

In actuality the often-used arguments for rejecting a kinetic-based determination in the past have collapsed with the availability of reliable instrumentation and easier data-handling approaches (see Chapter 1).

Great impetus in the development of instrumentation for kinetic methods was observed in the mid-1960s through the mid-1970s and can be traced back to the "operational amplifier revolution" (4). However, it was in the 1950s that a great deal of activity was dedicated to the development of recording spectrophotometers. The capability of continuous recording of a signal related to concentration changes as a chemical system evolves to equilibrium was, as indicated in Chapter 1, undoubtedly the first significant step in the instrumental promotion of measurements under dynamic conditions. Modular instrumentation received renewed impetus in the 1960s together with the strip chart potentiometric recorder and signal conditioning modules.

Table 1.1 summarized developments considered of interest in the evolutionary process by which instrumentation and computer applications have made kinetic measurements competitive with (and sometimes superior to) equilibrium-based measurements. Inspection of the table shows the evolution from the use of servo units of electromechanical nature to all-electronic systems.

This chapter presents the different means of mixing, the most common means of measurement, and some electronic signal conditioning and modification circuits and computer applications.

### 9.1. MEANS OF MIXING REACTANTS

The rates of slow chemical reactions in solution can be followed by simple and conventional methods of manual mixing, magnetic stirring, or both. The speed of initial mixing of the reaction components places a limit on the minimum half-time that can be handled in this manner. The mixing time with magnetic stirring bars, for instance, is a few seconds, so reactions with half-times smaller than 10 s are difficult to study with this type of mixing. For reactions that are over in 10 s or less, stopped-flow mixing offers an attractive alternative, albeit at the expense of simplicity.

#### 9.1.1. Mixing by Manual and Magnetic Stirring

Special reaction-mixture containers have been proposed and used in manual mixing for kinetic-based measurements. Figure 9.1a illustrates one attributed

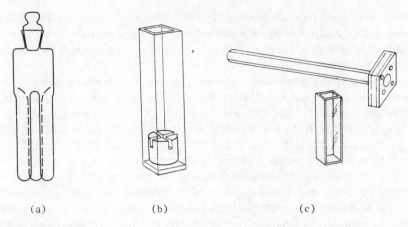

        (a)                       (b)                       (c)

**Figure 9.1.** Devices for aiding batch mixing. (*a*) Reaction-mixture vessel with individual compartments for up to three solutions. (*b*) Magnetic stirring button (Conrad design). (*c*) Cell mixer, plunger type.

to L. I. Budarin (5). The container has three compartments for three different solutions and is generally immersed in a constant-temperature bath to reach the desired temperature. When this is attained, the vessel is turned upside down and the contents energetically mixed. The resulting solution can then be transferred to a cell in which the measurement is performed, or aliquots can be withdrawn from the vessel at given times for measurement.

Convenient mixing can be accomplished directly in conventional cells for spectrophotometric monitoring by means of a magnetic stirring button designed by Conrad (6) (Figure 9.1*b*) or by means of a plunger-type mixer (Figure 9.1*c*), both of which are commercially available. These devices are generally used with syringe injection of one of the reactants directly into the cell; such injection is occasionally also used in continuous-flow mixing. The effciency of mixing in systems in which the force of injection is the sole mode of mixing has been addressed by Carter and Stanbridge (7). They injected one of the reactants (with a mechanically actuated syringe) into the other reactant(s) contained in a spectrophotometric cell. The rate at which iodide is oxidized by hydrogen peroxide (as measured by the time to consume a given amount of thiosulfate) was used as a measure of mixing efficiency.

### 9.1.2. Stopped-Flow Mixing

The same approaches used in kinetic-based determinations with relatively slow reactions can be applied to fast reactions, provided reliable fast mixing is available. Most reaction rate methods taking advantage of reactions with half-times of less than 10 s utilize the stopped-flow approach for mixing reactants.

The stopped-flow method employs a pair of driven syringes to force the reactants into a mixing chamber and then into the observation cell. As soon as the mixed solution reaches the cell, the flow is stopped and changes in the measured parameter (most of the time, absorbance change) are observed without interference from artifacts arising from flow and turbulence.

Figure 9.2 compares the mixing capabilities of conventional and stopped-flow mixing in terms of half-time of reaction and rate constants.

A typical stopped-flow setup is illustrated in Figure 9.3. The stopped-flow technique requires only about 100 to 500 $\mu$L of solution for a complete run, has dead times as low as 0.5 ms, and permits extending the observations to minutes, but if requires fast detection and rapid readouts.

Stopped-flow methods require a very rapid and efficient mixing of separate reagent streams that will constitute the final solution. A typical *mixing jet* is illustrated in Figure 9.3. Tangential ports in the mixing jet create turbulent flow that intimately and almost instantly mixes the two reactant solutions, which are forced into the mixing jet by pneumatically actuated drive syringes. The freshly produced mixture enters the observation chamber, forcing the spent fluid from the previous experiment out and into a stop syringe. When this syringe plunger reaches the mechanical stop, all flow is instantaneously stopped and a trigger switch initiates data collection by an oscilloscope, strip chart recorder, or computer. At the end of the measurement, the stop syringe is manually purged, making the instrument ready for the next experiment. The reactants are stored in the reservoir syringes and transferred, as needed, to the drive syringes by means of four rotary valves, which eliminate the possibility of cross contamination during filling.

Stopped-flow instrumentation can be rather simple; Harvey (8) has described an instrument simple enough to be used for teaching purposes, yet

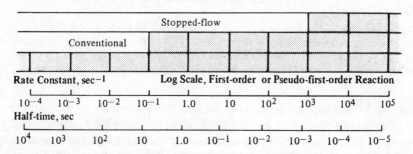

**Figure 9.2.** Comparison of the mixing capabilities of methods of conventional (batch) mixing and stopped-flow mixing. [Reproduced with permission of Dionex Corporation, Sunnyvale, CA.]

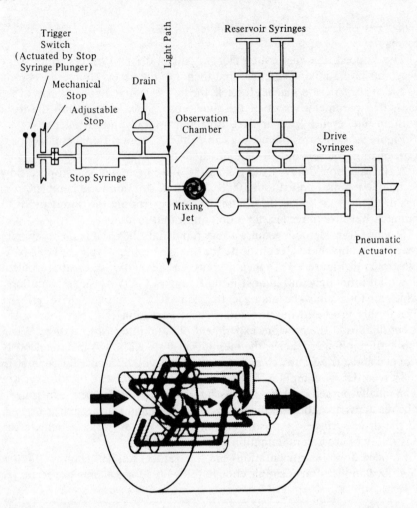

**Figure 9.3.** Schematic diagram of a typical experimental setup for stopped-flow mixing with photometric detection and details of the tangential mixing jet. Separate reagent streams enter from the left, and the (mixed) solution exits to the right to the observation chamber. [Reproduced with permission of Dionex Corporation, Sunnyvale, CA.]

sufficiently versatile to serve as a nucleus for a research instrument. The mixing system, illustrated in Figure 9.4, was inspired by a design originally proposed by Strittmatter (9). Operation of the mixing device can be summarized as follows. The solutions to be mixed are manually driven fast from the storage syringes, with the help of the pushing block, into a mixer situated in the base of a 3-mm-square observation cell. The reaction solution traverses

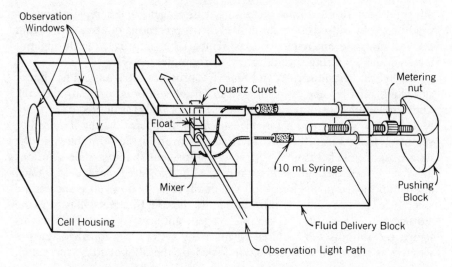

**Figure 9.4.** Harvey's fluid delivery system. For clarity the fluid delivery block is shown separated from the cell housing. In operation these modules interlock to form a light-tight compartment. [From Ref. 8, reproduced with permission of Academic Press.]

the mixing chamber and enters the observation cell, where it displaces a float that fits into the cell. The mixer, designed by Strittmatter, consists of two horizontal, mutually perpendicular channels (0.5-mm diameter) that enter a vertical mixing chamber (0.65-mm diameter × 2-mm height). After each determination the reaction mixture is displaced from the lower portion of the cell by depressing the float with the end of a blunt, wide-gauge needle.

The interested reader can obtain more information on the development of stopped-flow mixing in works preceding the publication of Harvey's paper (10–12). Crouch et al. (13) have discussed automated stopped-flow systems in a very comprehensive manner. Their review introduces the reader to advantages and limitations of reaction rate methods, examines in detail stopped-flow system components and their functions, and closes with analytical applications of stopped-flow mixing with spectrophotometric detection.

An interesting hybrid making use of an automated stopped-flow unit in conjunction with an unsegmented solution-storage unit has been proposed by Malmstadt et al. (14). The stopped-flow component is a relatively simple and compact microcomputer-controlled setup previously described by the same authors (15). The solution-storage component was inspired by developments in the technology for unsegmented continuous-flow sample–reagent(s) processing (unsegmented continuous-flow mixing is considered in Section 9.1.3). The marriage of these two types of solution handling retains the advantages of stopped-flow processing (e.g., reagent and sample are measured precisely and mixed quantitatively with precisions of about 0.1%, and fast reactions are amenable to measurement), but improves the throughput for reactions requiring several seconds or minutes.

The hybrid unit, illustrated in Figure 9.5, provides rapid automatic aliquot sampling, mixing, and sequential transfer of solution to a storage coil. Different lengths of storage coil can be used to provide the desired delay times and sample conditioning before measurement. In operation, the standard or sample is loaded into the syringe-drive system from the turntable, and the reagent(s) are loaded from a reservoir. The reagent(s) and sample solutions are then forced through a mixer into the thermostated storage coil. Each segment is moved through this coil and finally into the flow cell as succeeding sample–reagent aliquots are introduced into the coil. Figure 9.6 illustrates the storage within a coil for six sample–reagent aliquots. With the coil in this figure, six aliquots are required to flush the system. the fourth of the six aliquots is shown in Figure 9.6a, followed by the aliquot to be measured,

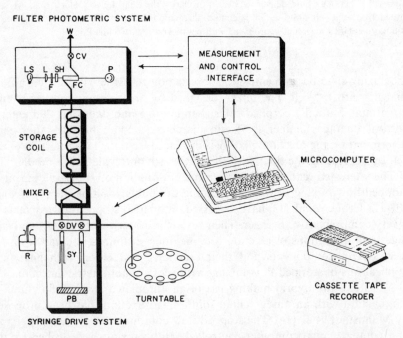

**Figure 9.5.** Block diagram of the unsegmented stopped-flow storage unit. LS, light source; L, lens; F, filter; SH, shutter; FC, flow cell; CV, check valve; W, waste; P, photomultiplier; R, reagent; DV, dual three-way flow valves; PB, plunger block; SY, syringes. [From *American Laboratory*, Vol. 12, No. 9, page 27. Copyright 1980 by International Scientific Communications, Inc.]

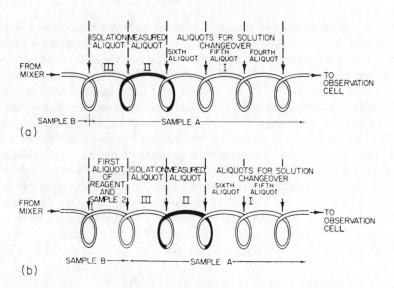

**Figure 9.6.** Schematic of the six-aliquot storage coil. (*a*) Coil showing three sections: I, aliquots needed for solution changeover; II, measured aliquot; III, aliquot following measured aliquot, which serves to isolate measured aliquot from next sample. (*b*) Coil shown 1.6 s later, after injection of next aliquot. [From *American Laboratory*, Vol. 12, No. 9, page 27. Copyright 1980 by International Communications, Inc.]

which is free of contamination from the previous sample. The next aliquot isolates the aliquot to be monitored from the succeeding sample. Figure 9.6*b* shows the contents of the coil one injection later. The first aliquot containing the next sample has now entered the coil; six more aliquots are then injected to avoid contamination from the preceding sample.

Each cycle in this system takes 1.6 s, during which aliquots of the sample and reagent(s) solutions are mixed and introduced into the storage coil. Consequently it will take 9.6 s for a given aliquot to traverse the length of a six-aliquot storage coil. Different delay times can be selected by changing the length of the coil.

Temperature effects are critical in stopped-flow mixing as a consequence of the friction generated in forcing a solution at high velocity through narrow-bore tubing. Holler et al. (16) performed a study of temperature effects in three different stopped-flow mixing systems. Two of the units evaluated are commercially marketed (Durrum Model 110 stopped-flow spectrophotometer and GCA McPherson Model Eu-730-11 stopped-flow module); the third was developed in the laboratory of the authors. the custom-made module exhibited large temperature variations from ambient ones, and the Durrum unit exhibited the least such variations in the 20 to 30°C range. The EU-730-11 also exhibited large variations from ambient temperature, but separate thermostating of the drive syringes and modification of the push cycle produced substantial improvements. The effect of temperature-induced refractive index gradients on the optical response of systems using stopped-flow mixing has been examined by Stewart (17). The author provides a quantitative explanation of these effects and suggests an optical means of minimizing them that could be applied in spectrophotometric measurements after stopped-flow mixing.

Worth mentioning among recent developments in the area of stopped-flow instrumentation are a microcomputer-automated on-line reagent dilution system reported by Stieg and Nieman (18) that significantly reduces the time for reagent(s) preparation, and an automated commercially available stop-ped-flow spectrophotometer using two microcomputers (19). This parallel processing system is a simple and inexpensive way of overcoming limitations associated with the low speed of microcomputers for the extensive comput-ations required in some reaction rate methods.

### 9.1.3.  Continuous-Flow Mixing

There are at least three distinctive areas in which continuous-flow mixing is used in kinetic-based methodology: (a) the study of fast reactions, (b) air-segmented sample processing, and (c) unsegmented sample processing.

In the continuous-flow method for the study of fast reactions (20), the reactants flow in separate continuous streams that meet in a mixing chamber and then pass along an observation tube or chamber with detection devices placed at appropriate points along its length. Reactions with half-times in the neighborhood of 0.001 s can be studied by this form of mixing and solution handling. This type of mixing allows the use of slowly responding detectors because at any particular point in the flow stream the "age" of the solution is constant; however, continuous-flow mixing requires large reaction volumes, a disadvantage when compared with stopped-flow mixing. With dead times as low as 100 to 200 $\mu$s, continuous-flow operation is slightly faster than stopped flow because mechanical problems associated with the abrupt stopping of the flow are absent. However, this type of mixing has found very limited used for analyte determinations.

Air-segmented and unsegmented continuous-flow sample–reagent(s) pro-cessing developed primarily as an analytical answer to the increasing load being imposed on the practicing chemical analyst (21). These methods apply mainly to the problems posed when a large number of samples of similar nature have to be processed for the determination of a single species. Such problems are common to clinical chemistry, environmental studies, industrial processes, and quality control. Short historical overviews of early develop-ments can be found for both air-segmented (22) and unsegmented (21, 23) continuous-flow sample–reagent(s) processing. Air-segmented systems preceded unsegmented ones, and Skeggs was responsible for their introduction in the analytical (clinical) laboratory (24). Skeggs's implementation of air-segmented continuous-flow analyses drastically changed the way colorimetric determinations (and by extension practically all instrumental variations in wet chemistry) were performed. The most radical contribution, however, was the introduction of measurements under dynamic conditions instead of the classical "static" measurements using a cell or cuvette. It is to be noted, however, that the elements of the modular, continuous-flow concept existed in process control before Skeggs adapted them to the analytical laboratory (21).

Mixing in air-segmented continuous-flow systems is illustrated in Figure 9.7. Samples and wash liquid are intermittently aspirated at the sampler. The resulting stream is segmented by air bubbles and combined, for instance, with a diluent A, passed through a mixing coil and a dialyzer. Although not always necessary, the dialyzer became a standard unit in the the commercial versions introduced by Technicon Corp. around 1957. Dialyzable analyte(s) diffuse through the dialysis membrane and into a second segmented stream B, combines with another reagent C, and passes through a thermostated reaction coil in which a chromophore is developed or removed from the system. The final step is passage through the detection area (colorimeter). It is obvious that other means of detection can be coupled to these mixing systems and that the extent of chemical reactions without color changes can be

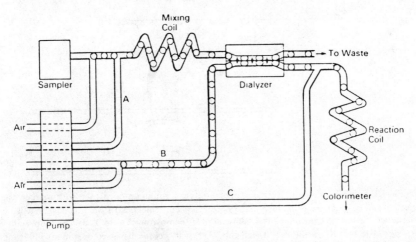

**Figure 9.7.** Air-segmented continuous-flow sample–reagent(s) mixing. A, B, and C may be reagent(s) or diluent lines. [Reprinted with permission from L. Snyder et al., *Anal. Chem.*, **48**, 942A. Copyright 1976 by the American Chemical Society.]

monitored equally well by means other than colorimetry. Direct sample transport to detection (i.e., with no chemical transformations imposed on any of the sample components) is also equally possible and has been implemented with both air- and unsegmented continuous-flow systems.

Mixing in this type of continuous-flow sample–reagent(s) processing takes advantage of the natural motions of fluids that occur within a short segment of liquid enclosed at each end by a gas–liquid interface. The resulting fluid motion (*bolus flow*) is illustrated in Figure 9.8. This type of mixing requires no external means such as stirrers or vibrators; sufficient length of tubing (mixing time) is all that is required. Bolus mixing results in rapid longitudinal exchange of fluid within short segments. The rate-limiting mixing process occurs radially across the fluid streamlines. In the straight portions of the tubing manifold, radial mass transfer is mainly the result of molecular diffusion. Such diffusion is rather slow in liquids, and helical-coiled portions are employed to generate convective radial mixing.

Studies of mixing efficiency (25) have shown that viscosity, density, and flow rate of liquids affect their mixing rates; viscosity is the physical property with most significant effect. Tube internal diameter, helix coil diameter, and liquid-segment length also affect mixing. The time distribution (concentration profile) of the monitored species is altered by the axial dispersion of analyte molecules along the flowing stream. Excessive dispersion results in *sample*

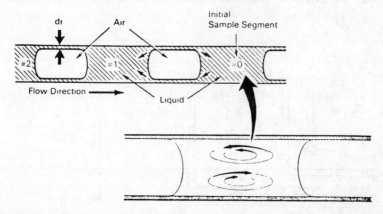

**Figure 9.8.** Sample dispersion during segmented flow through an open tube with marker dye injected into segment 0. Film of liquid phase ($d_f$ = film thickness) from segment 0 will be overtaken and mixed into segment 1, allowing dye to move from 0 to 1; dye buildup within segment 1 can also be transferred to segment 2. Dye will become dispersed over many following segments. Lower detail shows the bolus circulation within a moving liquid segment, which occurs naturally in any segmented-flow system. [Reprinted with permission from L. Snyder et al., *Anal. Chem.*, **48**, 942A. Copyright 1976 by the American Chemical Society.]

*carryover* (invasion of sample components into adjacent liquid segments), a problem that is minimized by air segmentation and by increasing the inter-sample wash time at the expense of sampling rates. The physical basis of the dispersion process can be seen with the aid of Figure 9.8. A dye used as a tag is injected into segment 0. Since the flowing stream wets the inside walls of the tubin, a thin film of dye is laid down by each moving segment. The film from segment 0 is thus overtaken and mixed into segment 1. This allows some dye to move from segment 0 to segment 1. The buildup of dye in segment 1 is subsequently transferred to segment 2, and so on. In this way the dye originally contained only in segment 0 is dispersed over several following segments. Effectively designed mixing manifolds result in minimal dispersion. Further details of basic as well as applied principles of air-segmented systems can be found in appropriate monographs (26).

Mixing in unsegmented streams by introducing a sample plug into a continuously flowing carrier has a rich developmental background. It can be traced back, for instance, to experiments aimed at estimating the quantity of blood put out by the heart and lungs (indicator dilution technique) (27) and to analytical methodology for continuous measurement of enzymatic reaction rates by Blaedel and Hicks (28). Figure 9.9 illustrates the flow system proposed by Blaedel and Hicks, which permits fixed-time determinations as well as continuous measurement of reaction rates. The absorbance of the reacting mixture is first recorded at the upstream cell and, after a fixed time delay, again at the downstream cell. The ratio of absorbances in the cells and the travel time between the two cells provide the data for calculations.

Unsegmented continuous-flow sample–reagent(s) processing did not attract wide attention and use until the mid-1970s when Ruzicka and Hansen coined the term *flow injection analysis* (29) and Stewart et al. (30) described a high-speed discrete-sample analyzer based on unsegmented continuous flow. Since then, unsegmented systems have found a myriad of analytical uses and applications, and monographs on the subject have been published (31, 32). Figure 9.10 illustrates the basic components of the simplest flow injection manifold. The approach may vary somewhat, but a common characteristic of procedures using unsegmented continuous flow, as the name implies, is the introduction of the sample by a discrete injection (better described as *intercalation* when a sliding valve is used) into a continuously flowing, unsegmented stream. The carrier stream transports the sample "plug" to the detection area in which the analytical signal (measurement) is acquired. During sample transport, chemical reactions between the species to be determined and chemical components of the carrier stream may occur and provide the basis for unsegmented mixing. It should be indicated, however, that unsegmented continuous flow can and has been used solely as a means of sample transport (e.g., in the monitoring of fluoride ions reported in Ref. 33). In any case,

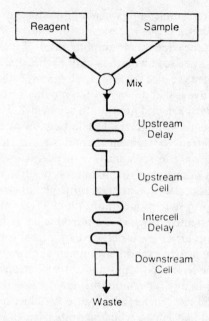

**Figure 9.9.** Schematic diagram of instrumental setup for fixed-time enzymatic determinations and for continuous measurements of reaction rates. [Reprinted with permission from W. J. Blaedel and G. P. Hicks, *Anal. Chem.*, **34**, 388. Copyright 1962 by the American Chemical Society.]

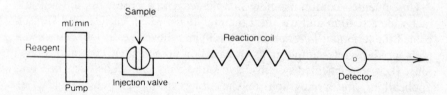

**Figure 9.10.** Simplest form of an unsegmented-flow injection manifold.

detection occurs in the unsegmented continuous-flow stream while the system is attaining equilibrium by a physical or a chemical process or both. Thus in essence, unsegmented continuous-flow procedures are kinetic-based. Because the flow injection approach makes measurements under dynamic conditions rather simply and inexpensively, its popularization has been explosive in the past decade. This was evident in the number of papers presented at the First International Symposium on Kinetics in Analytical Chemistry (34) and the Second International Symposium on the same topic (35) dealing with applications of unsegmented continuous-flow mixing.

As with segmented continuous flow, mixing in unsegmented systems results from a combination of diffusion and convection within and around the boundaries of the injected sample plug under predominantly laminar-flow conditions. Also as in segmented manifolds, helically coiled tubings are used to enhance radial mixing (secondary flow phenomena), or there is incorporation of devices such as the so-called single-bead-string reactor (36) illustrated in Figure 9.11. The beads, whose diameter is about 65% of the diameter of the tubing, induce eddy movements as the plug travels through the reactor. The overall effect is enhanced mixing, with characteristics similar to an increase in diffusional mixing. The effect of plug dispersion as it travels from point of injection to point of detection is illustrated in Figure 9.12. The predominance of diffusion or convection depends on the flow rate, the radius

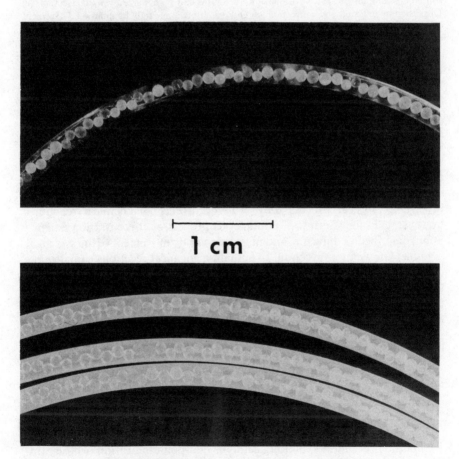

**Figure 9.11.** Photograph of single-bead-string reactors.

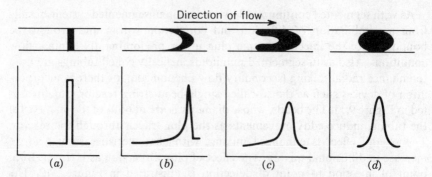

**Figure 9.12.** Representation of concentration profiles and shape of sample plug in unsegmented continuous-flow mixing. (*a*) No dispersion. (*b*) Dispersion predominantly by convection, (*c*) Dispersion by diffusion and convection. (*d*) Dispersion predominantly by diffusion. [Reprinted with permission from D. Betteridge, *Anal. Chem.*, **50**, 832A. Copyright 1978 by the American Chemical Society.]

of the tubing, the residence time of the plug in the manifold, and the diffusion coefficients of the several species in the medium.

The topic of dispersion in unsegmented continuous-flow mixing has been reviewed in detail (38, 39). In its physicochemical context the word *dispersion* characterizes the distribution (scattering) of matter or radiation in a dispersing medium; a typical example is a colloidal system consisting of a dispersed phase in a dispersion medium. Although many colloidal systems are the result of chemical reactions, the term *dispersion* is commonly considered to involve a physical process only. The occurrence of chemical reactions, however, can alter both the spontaneous process of concentration equalization, described as diffusion, and mass transfer by movement of the system as a whole, known as convection. In actuality, chemical reactions can alter the concentration profile, which in turn leads to changed diffusion and convection; these then affect the collisional frequency that governs chemical reactions. This interdependence between physical and chemical processes defines the complexity of mixing in unsegmented flows when chemical reactions alter the physical picture of dispersion (40, 41). Details on the kinetics of continuous-flow mixing can be seen in Chapter 11.

An attractive feature of unsegmented mixing is its amenability to miniaturization (42, 43).

### 9.1.4.  Mixing for Kinetic-Based Titrimetry and Stat Procedures

Titrimetric procedures exploiting catalytic end-point indication (whose principles were discussed in Chapter 4) are typical examples of experimental procedures requiring a uniform, accurate, reproducible, and continuous infusion of solution (titrant).

Motor-driven burets and syringe pumps are particularly useful for delivering a titrant into a vessel external to the detection cell (44) or directly into the cell (45) with magnetic mixing. A titration head based on a motor-driven buret of variable and reproducible rate of titrant delivery and a special cell holder involving an air-driven stirrer have been described by Hall et al. (45). Figure 9.13 shows details of the titration head and switching circuit for operation of the syringe delivery system. This system, is of interest for delivery of rather concentrated titrants because it allows addition of small volumes and avoids corrections for dilution. Its design allows one to change rather easily the rate or total volume introduced by selecting motor, syringe, or both. This makes the system of rather general utility. The switching circuit was designed to permit an almost unattended and safe operation.

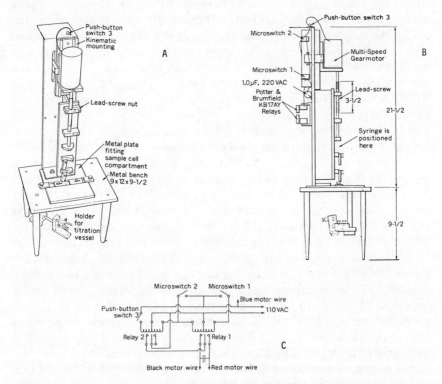

**Figure 9.13.** Titration head for uniform, accurate, reproducible, and continuous delivery of titrant. (*A*) Perspective—frontal view. (*B*) Side view. (*C*) Switching circuit. All measurements in inches. [From Ref. 45, reproduced with permission of Academic Press.]

In the driving position the lead-screw is advanced downward, pushing the plunger and delivering the liquid. When the plunger has reached a point close to the end of the syringe, an adjustable screw actuates relay 1; this immediately reverses the direction of the motor, raising the lead-screw automatically. The motor, which slides on a kinematic mounting, pulls up the lead-screw. When it reaches the top, it is stopped by relay 2 and microswitch 2. As the motor and lead-screw move up, the plunger disengages from the plastic piece, which pushes it down, allowing very easy removal for either refilling or cleaning. The return of the lead-screw to the upper position can be speeded up by substituting a split nut for the lead-screw nut in the setup. Depressing push-button switch 3 initiates delivery again.

Stat procedures are an example of kinetic-based determinations centered on the delivery of a solution containing a chemical species capable of counteracting a chemical change in the system. If the rate of delivery matches the rate of change in the system, a steady-state situation is approached (stat condition). The approach originated in the determination of enzyme activity with the maintenance of pH at about a constant value (generally the optimum pH for the enzyme action as a catalyst) by a pH-stat procedure. The rate of base or acid addition parallels the rate of chemical change in the system. The principles and instrumentation for pH-stat determinations have been reviewed (46), and other stat procedures have been implemented for kinetic-based determinations (e.g., absorptiostat (47), biamperostat (48), and luminostat (49)).

A continuous-flow stat method for monitoring sample streams within certain limits of concentration has been proposed by Weisz and Fritz (47). The components of the absorptiostat setup are illustrated in Figure 9.14. Figure 9.15 shows the circuit diagram of the control system. Reactants 1 and 2 are separately pumped (channels R1 and R2) into the mixing chamber (MC). A third reaction component (e.g., a catalyst) is also pumped into the mixing chamber (channel R3). After mixing, the solution flows into the vertical cuvet (C). Light from the light pipe (LP) enters the cuvet and reaches the photodetector (P). The voltage developed at the detector is transferred to the control system and is recorded (vs. time) on one channel of a two-channel recorder. The command voltage is recorded in the other channel and is converted by the motor (M), which is mechanically coupled to the 10-turn potentiometer (see Figure 9.15). This potentiometer controls the pumping rate of the peristaltic pump. Changes in concentration are appropriately paralleled by an increase or decrease of the pumping rate.

It is of interest that sampling and mixing in continuous-flow systems also play an important role in process analyzers (50) and that conceptual situations in this field are relevant to continuous-flow mixing in analytical chemistry.

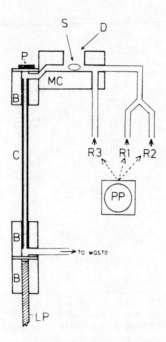

**Figure 9.14.** Continuous-flow absorptiostat. B, Plexiglass blocks; C, vertical glass cell; D, opening for debubbling; LP, fiberglass pipe connecting light sources to observation cell; MC, mixing chamber; P, photodetector; PP, regulated peristaltic pump that delivers the solutions to one of three channels (R1, R2, or R3). The unregulated peristaltic pump for the remaining two channels is not shown. [From Ref. 47, reproduced with permission of Elsevier Science Publishers, B. V.]

### 9.1.5.  Centrifugal Mixing

Centrifugal mixing is used in parallel fast analyzers. The principle of centrifugal mixing can be illustrated with the help of Figure 9.16 (51). The figure shows the situation for a sample and a single reagent, but there is no reason to limit to two the number of solutions to be mixed, since more than two wells could be molded into each slot of the disk. The disk is placed in a centrifuge rotor. When the rotor spins, mixing is induced by the action of the centrifugal force, which finally displaces the mixed solution into the observation cell (labeled "reaction mixture after transfer" in the figure). A second means for rapid mixing is provided by small siphons that connect the sloping outer edge (or bottom) of each well. Air or another gas may be forced through these siphons if the rotor is closed; the net effect is very rapid mixing with a sudden burst of bubbles. Centrifuge speeds of about 600 rpm are common during the

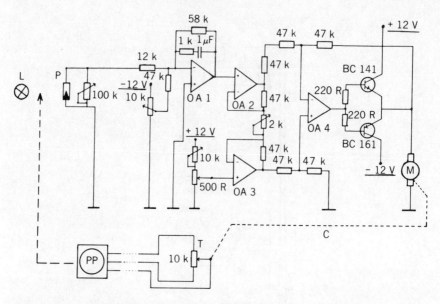

**Figure 9.15.** Circuit diagram for the control unit designed for the flow absorptiostat illustrated in Figure 9.14. C, Mechanical coupling; L, lamp; M, motor, mechanically coupled to a 10-turn potentiometer, T; P, photodetector; PP, regulated peristaltic pump. [From Ref. 47, reproduced with permission of Elsevier Science Publishers, B. V.]

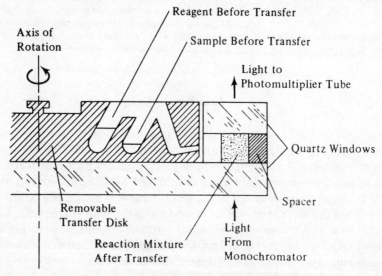

**Figure 9.16.** Cross section of the sample disk and centrifugal analyzer section of a parallel fast analyzer. [From *American Laboratory*, Volume 3, No. 7, page 26. Copyright 1971 by International Scientific Communications, Inc. Reproduced by permission.]

data collection portion of the operation but initial speeds can be in the range of 1000 to 2000 rpm.

Centrifugal mixing was pioneered by N. G. Anderson. He has provided an account of this form of mixing and data collection (52), which developed as part of the Fast Analyzer Project implemented within the MAN (Molecular Anatomy) Program at Oak Ridge National Laboratory. Efforts to miniaturize this kind of mixing have been discussed by Scott and Burtis (53).

## 9.2. DETECTION APPROACHES

Optical methods based on absorptiometric (spectrophotometric) detection outnumber all other forms of detection in the literature devoted to kinetic-based methods. Photon-emitting processes (e.g., luminescence such as fluorescence and chemiluminescence) have found increasing applications in recent years but are far from reaching the level of use of absorptiometric or electrochemical detection. Electrochemical detection, primarily by amperometry and potentiometry, is second in order of popularity, with some other detection approaches (e.g., thermometry and polarimetry) as distant followers.

### 9.2.1. Absorptiometric Detection

The design and construction of a spectrophotometric system particularly developed for kinetic-based determinations was dealt with in detail by Weichselbaum et al. (54, 55). Two key design criteria for the construction were that (a) the instrument should be very stable (noise-free), and (b) the instrument should be capable of giving a direct readout of the derivative of the absorbance–time curve. The second criterion arose because many analytical applications call for "initial rate" measurements, and the derivative readout gives an estimate of the reaction rate as a function of time.

The basic diagram for the derivative circuit is shown in Figure 9.17 (54). The derivative of absorbance with respect to time can be given as

$$d(\log T)/dt = (-\log e)(1/T)(dT/dt) \tag{9.1}$$

which can be obtained electrically by differentiating the transmittance and reading the result with a servo system with transmittance as a reference that provides a true electrical analog of eq. 9.1. The servo acts as a null seeker. The servo potentiometer is driven to equal the differentiator output, and the potentiometer setting is read on a mechanical counter dial.

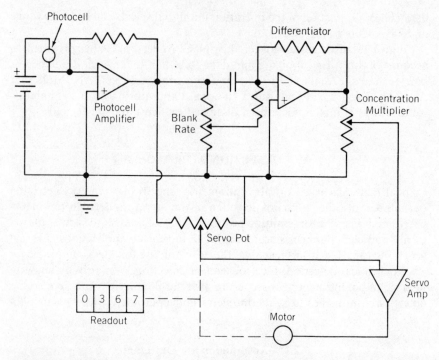

**Figure 9.17.** Basic diagram of the absorbance derivative circuit. [Reprinted with permission from T. E. Weichselbaum et al., *Anal. Chem.*, **41**(3), 103A. Copyright 1969 by the American Chemical Society.]

High-frequency noise can be minimized by simple electronic filtering; frequency of the noise is generally very high compared to the rate of signal change. Low-frequency noise and drift are more difficult problems to handle; they require light-source stabilization and minimization of detector noise and drift.

Noise minimization in photometric units to be used for reaction rate measurements depends greatly on light-source stabilization since these instruments are commonly of the single-beam design. Figure 9.18 illustrates the circuit employed by Weichselbaum et al. (55) for source stabilization. The voltage to the inverting input of the amplifier is the sum of the voltages across resistors $R_2$ and $R_3$. The voltage across $R_2$ is a fraction of the voltage across the bulb, and the voltage across $R_3$ should be proportional to the current through the light bulb. Variations of the ratios of resistors allow adjustment of the weight given to current and voltage feedback to compensate for variations in the light-bulb resistance. A baffled lamp is used to minimize variations arising from heat convection flow around the light bulb. A

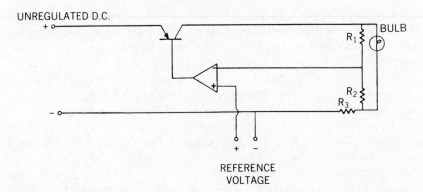

**Figure 9.18.** Schematic diagram of regulated lamp power supply circuit. [Reprinted with permission from T. E. Weichselbaum et al., *Anal. Chem.*, **41**, 725. Copyright 1969 by the American Chemical Society.]

photometric system with very high stability for extended periods of time has been proposed by Pardue and Rodriguez (56). The heart of the stabilization in this system is illustrated in Figure 9.19. The system utilizes optical feedback to maintain light-source stability. Long-term stability is rated at $\pm 0.02\%$ $T$ over several hours and $\pm 0.01\%$ within the hour. A beam splitter sends one portion of the light beam through the sample compartment to a phototube

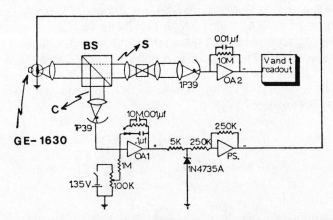

**Figure 9.19.** Schematic diagram of high-stability photometric system with optical feedback. The light beam from the tungsten lamp source (GE-1630) passes through a beam splitter (BS), which divides it equally between the sample and control beams (S and C respectively). [Reprinted with permission from H. L. Pardue and P. A. Rodriguez, *Anal. Chem.*, **39**, 901. Copyright 1967 by the American Chemical Society.]

and a second portion of the same beam to a control phototube. The output from the control phototube, after amplification, drives a programmable power supply, which in turn drives the radiant energy source. This feedback compensates for variations in the radiant power delivered by the source.

Detector noise and drift comprise the other area of concern in dealing with instrument stability. When the operator can control the light level, it is recommended that intense light sources be used with a phototube detector (57). The use of *fast optics* (optical systems characterized by low aperture ratios) offers the possibility of high light throughput. For systems where the light level cannot be controlled (emission of radiant energy such as in chemiluminescence measurements) or where the light level is very low, photomultiplier tubes give better performance than phototubes. The stability of the power supply used for the detector is more critical in the use of photomultipliers than in the use of phototubes since the gain is highly dependent on the power-supply voltage. The stability of the power supply should be at least an order of magnitude greater than the desired stability gain.

The use of photodiode detector–amplifier combinations are of special interest because they offer a flat noise spectrum to DC without hysteresis or memory effects as a characteristic of spectral stability. These systems normally combine high-reliability PIN silicone photodiodes and low-current-mode operational amplifiers in a single compact package. They commonly use FET technology and hybrid packaging to increase signal-to-noise ratios. They provide the advantages of photomultipliers in the form of solid-state devices requiring low-voltage power supplies (e.g., $\pm 10$ to $\pm 20$ V) and presenting a relatively large (typically 1 cm$^2$) area. A PIN diode is a silicon semiconductor consisting of a layer of intrinsic (high-resistivity) material contained between highly doped $p$- and $n$-type materials. Figure 9.20 illustrates a typical circuit for a PIN/OPAMP combination. The output voltage of units based on the circuit of Figure 9.20 is linearly related to the input radiant power by the following formulation:

$$e_{out} = P_1 R R_f \tag{9.2}$$

where $P_1 =$ the light power (in microwatts), $R =$ the responsivity of the photodiode (in microamperes per microwatt), and $R_f =$ the feedback resistance (in ohms).

A detection approach that has changed the perspective of UV-visible spectrophotometry is the use of photodiode arrays (58, 59). This multichannel detection has brought the possibility of measuring a full-wavelength-range spectrum in less than 1 s; its impact on kinetic methods has been reviewed by Valcárcel (60).

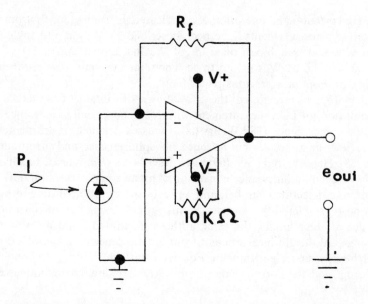

**Figure 9.20.** Typical circuit for a PIN diode–operational amplifier combination. Low light levels generally require feedback resistors of at least 500 kΩ; 100 Ω to 1 kΩ are used with high light levels. The variable 10-kΩ resistor is used for zero offset. Potential $V$ is typically ±15 V.

### 9.2.2. Detection by Means of Fluorescence, Chemiluminescence, and Bioluminescence

As mentioned in the preceding chapter, fluorescence monitoring provides advantages such as high sensitivity, low limits of detection, and relatively large dynamic ranges in calibration plots. Instrumental considerations for fluorescence monitoring have been addressed by Ingle and Ryan (61). Most of the requirements are common to other optical methods of monitoring the progress of a reaction, and a review of fluorometric reaction rate methods for the determination of inorganic species illustrates the instrumental requirements concisely (62). Within the realm of photon emission and optical detection fall chemiluminescence and bioluminescence methods (63, 64). The measurements utilized in these applications are dynamic in nature, and as such they deserve a place in kinetic-based methods. Most analytical applications of luminescence require the measurement of either peak light intensity or integration of the area under all or part of the light emission–time curve. Photoelectric detectors, cells, mixing approaches, auxiliary electronics, instrumental standards, and instruments for chemiluminescence measurement have been discussed by Seitz in some detail (63).

The construction and operation of a dedicated instrument for fluorometric reaction rate measurements have been described by Wilson and Ingle (65). Figure 9.21 shows a block diagram of the instrument and a list of the components used by Wilson and Ingle. The ratemeter unit was described by the same authors in a previous paper (66).

Stanley (67) has reviewed the availability of suitable commercial instrumentation for light measurement with focus on chemiluminescence and bioluminescence. Since 1980 Wehry has discussed pertinent instrumentation in his reviews on molecular fluorescence, phosphorescence, and chemiluminescence (68). Hayashi and Yuki (69) have described a computerized analyzing system for chemiluminescence consisting of an automatic injection device, a reaction and detection chamber, and a processing unit. Rapid time response and good reproducibility characterize this system. Using luminol chemiluminescence as their model, the same authors illustrated rapid measurement involving peak-height measurement from luminescence data smoothed by a digital filter, a fast algorithm for Fourier transform filtering for real-time processing, and the corresponding application to slow chemiluminescence (70).

### 9.2.3.  Electrochemical Detection

The second largest group of measuring techniques that have been applied to kinetic-based methods is electrochemical in nature. From the viewpoint of convenience in monitoring the course of a reaction or process, electroanalytical sensing shares the advantages of detection of absorption and emission: it does not affect the reaction media, solutions can be readily thermostated, and the measured parameter can be continuously recorded.

The two types of electrochemical techniques most often used are potentiometry and amperometry. Impetus in the use of detectors in these areas, for measurement under dynamic conditions, has resulted mainly from an increase in the popularity of electrochemical detection in continuous-flow systems (e.g., liquid–liquid chromatography and flow injection analysis) (71). Potentiometric and amperometric detection in flow injection enzymatic determinations (monitoring of protons and dissolved oxygen levels) has been reviewed (72). In these detectors, which are surface-type and nonintegrating, the signal is dictated by the concentration or activity of a monitored species in the proximity of the sensing electrode. This sets them apart from bulk-property (e.g., conductometric) and optical (photometric) detectors. Detection requires transport of the electroactive (monitored) species from the bulk to the diffusion layer (in the vicinity of the working electrode) and across the diffusion layer to the sensing surface of the working electrode. As a result of this, surface sensors may respond to localized artifacts more easily than do

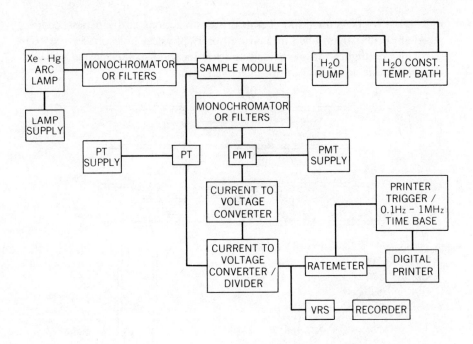

| Lamp power supply | Electro Powerpacs Corp., model 352 |
| Lamp housing | Schoeffel Instrument Corp., model LH150 |
| Monochromators ($F/n = 4.7$) | Schoeffel Instrument Corp., model GM100 |
| PMT housing | Heath-Schlumberger, model EU-701-93 |
| PMT supply | Heath-Schlumberger, model EU-42A |
| PT supply | Heath Universal Power Supply, model EUW-15 |
| Recorder | Heath-Schlumberger, model SR-255B |
| Magnetic stirrer | Waco Midget, model 86335 |
| Stirring bar | Bel-Art, model F37150 |
| Constant temperature circulator | Techne, model TU-12 |
| Pump | Jabsco, model P1-M6 |
| Hg–Xe lamp (200 W) | Hanovia, model 901B |
| PT | RCA, model 935 |
| PMT | RCA, model 1P28 |
| PMT OA | Function Modules, Inc., model 380J |
| PT OA | Analog Devices, model A 540J |
| Divider | Analog Devices, model 434B |
| Digital printer | Newport Laboratories, Inc., model 810 |
| Voltage reference source (VRS) | Heath, model EU-80A |
| Water bath | Pyrex, 12 in × 12 in jar |

PMT = photomultiplier; PT = phototube; OA = operational amplifier.

**Figure 9.21.** Block diagram and list of components for a dedicated fluorometric reaction rate instrument. From Ref. 65. Copyright 1977 by the American Chemical Society. Reproduced with permission.

bulk property detectors, and chemical (side) reactions and kinetics become more important than when measuring purely physical bulk properties. The particular configuration of the cell and the location of the sensing surface, then, need due consideration for effective sensing.

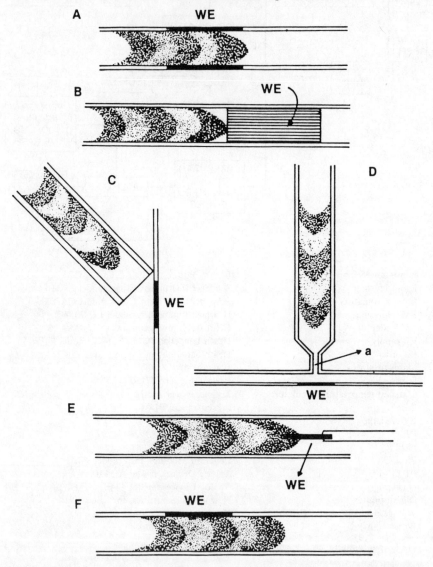

**Figure 9.22.** Basic configurations for electrochemical detectors used in continuous-flow sample processing. (*A*) Wall-tubular. (*B*) Porous (tubular). (*C*) Cascade. (*D*) Wall-jet. (*E*) Wire. (*F*) Thin-layer. WE, working electrode surface; a, Inlet nozzle. [From Ref. 72. Reproduced with permission of Elsevier Publishing Co].

Figure 9.22 depicts the most commonly used assemblies for both potentio-metric and amperometric detection in flow systems. Of the six configurations shown in the figure, the three most commonly employed in continuous-flow analyses are the wall-tubular, cascade, and wire types. The wall-jet and the thin-layer configurations have been employed mostly in chromatographic detectors, and the porous (tubular) configuration in electrochemical studies. The cascade type has been used in potentiometric sensing with ion-selective electrodes, including those based on field-effect transistors. The wire, wall-tubular, and thin-layer types have been used for amperometric detection. Except for the wall-jet type, the flow conditions predominant at the detection zone are laminar in all basic configurations; when the wall-jet type is used, laminar flow normally should prevail in the transporting tubing up to the point of detection.

Bipotentiometric monitoring has also found a place in kinetic-based methods. The potential difference between two electrodes (generally platinum) polarized by a small constant current is measured as a function of time. The approach is particularly suited to the monitoring of catalyzed reactions (including enzymatic ones). A difference in redox potentials for reactant(s) and products is a necessity in bipotentiometric monitoring. Differ-ential constant-current potentiometry defines a measurement in which the difference in potential between two electrodes polarized by different constant currents is monitored (73). Shifts in potential under stirring conditions occur when the concentration of electroactive species changes in the vicinity of the electrode. A peak-type signal is exhibited by an electroactive species undergo-ing depletion by homogeneous reaction in solution. The width of the peak is proportional to the homogeneous reaction rate. Because the type of measure-ment is differential, ordinary sources of error tend to cancel. The approach has been applied to the determination of phenols by bromination and to the oxidation of hydroxylamine by hexacyanoferrate(III) (73). Figure 9.23 shows a block diagram for differential constant-current potentiometric measure-ment.

Electrolytic conductance measurement is considered highly accurate and precise; however, its application to kinetic-based methods has been quite limited. A bipolar pulse modification, under computer control, of conduc-tometry has been developed by Caserta et al. (74) for use as a detection unit in reaction rate determinations, where its high sensitivity is an attractive feature.

### 9.2.4.  Other Methods

Enthalpimetric measurements (75) involve the change of a physical property (normally the resistance of a semiconductor device, such as a thermistor) as a result of heat evolution during the course of a chemical change. Direct-injection enthalpimetry (75) has proved to be particularly suited for applic-

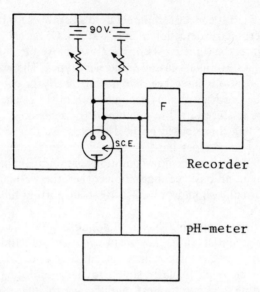

**Figure 9.23.** Differential constant-current potentiometric circuit. F, 1:1 voltage follower for impedance matching. [Reprinted with permission from J. B. Sand and C. D. Huber, Jr., *Anal. Chem.*, **42**, 238. Copyright 1970 by the American Chemical Society.]

ation to kinetic determinations. The enthalpimetric monitoring of catalytic end-point detection has been reviewed by Greenhow (76).

Jacobs et al. (77) have interfaced a rapid-scanning electron paramagnetic resonance spectrometer with a stopped-flow mixing unit capable of reaction rate monitoring. Raman spectrometric detection coupled to a microdroplet mixing unit has also been used for following the rates of rapid reactions (78).

## 9.3. ANCILLARY ELECTRONIC UNITS

The electronic boom that made such inroads in analytical instrumentation in the late 1950s and early 1960s greatly contributed to the acceptance of time-dependent measurements as a viable alternative to measurements made with systems at equilibrium. Thus efforts that evolved from automatic units to complete instrumentation (automated) systems (79) have received attention in several recommended reviews (79–81). A few specific examples are mentioned here.

### 9.3.1. Circuit for Automatic Measurement of Slopes of Rate Curves

An instrumental approach for the automatic measurement of the slope of the rate curve near zero reaction time was offered by Pardue back in 1964 (82).

The measurement entails matching the slope of the rate curve with the output from an electronic integrator. Figure 9.24 shows a schematic of the circuit. A servomechanism compares the signals from the chemical system and integrator and adjusts the integrator input until both slopes are the same. At balance the input to the integrator is proportional to the slope of the rate curve. The measurement step is usually rapid, requiring less than 20 s, and the rate data (slopes) can be displayed continuously by a strip chart recorder or read numerically from a meter. Other systems for automatic direct readout of rate data utilizing servocomparison techniques have been proposed by Malmstadt et al. (83, 84). One of these (84) affords measurement cycles that can be varied from milliseconds to hundreds of seconds and can provide optimum performance for very fast or slow reactions with very low relative errors and competitive precision.

### 9.3.2.   A Fixed-Time Digital Counting System

This system was devised to compute digitally the initial rate of a reaction by integration (85). As is generally the case, digital computation provides increased speed, accuracy, and reliability when compared with analog computation. Figure 9.25 gives a block diagram of the fixed-time digital computing unit, and Figure 9.26 shows the circuit of the fixed-time rate

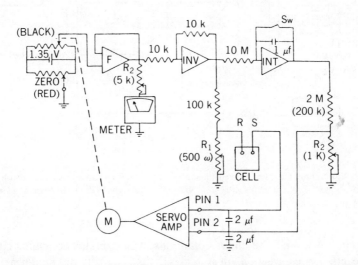

**Figure 9.24.** Circuit diagram of unit for automatic measurement of slopes of rate curves. F, follower; INV, inverter; INT, integrator. Cell corresponds to the case of potentiometrically following the rate of a reaction. [Reprinted with permission from H. L. Pardue, *Anal. Chem.*, **36**, 633. Copyright 1964 by the American Chemical Society.]

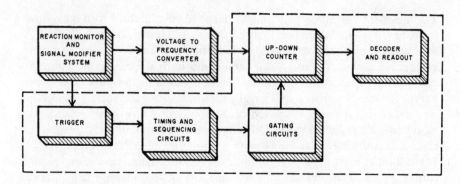

**Figure 9.25.** Block diagram of a system for reaction rate monitoring using fixed-time digital counting. Fixed-time rate computer shown in dotted lines. [Reprinted with permission from J. D. Ingle, Jr., and S. R. Crouch, *Anal. Chem.*, **42**, 1055. Copyright 1970 by the American Chemical Society.]

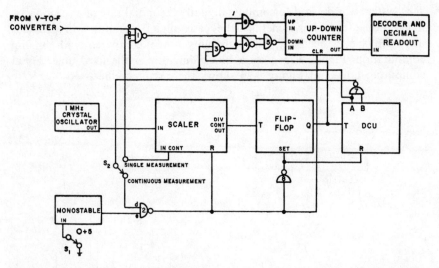

**Figure 9.26.** Circuit diagram for fixed-time rate computer. [Reproduced with permission from J. D. Ingle, Jr., and S. R. Crouch, *Anal. Chem.*, **42**, 1055. Copyright 1970 by the American Chemical Society.]

computer. Although the digital counting system contains no analog circuits, the reaction rate measurement system is limited by drifts and nonlinearities in the reaction monitor-signal modifier and in the voltage-to-frequency (V-to-F) converter. Replacement of these analog circuits with a photon-counting system (e.g., direct interfacing of the photomultiplier tube with a fixed-time rate computer) would yield an all-digital system with significant improvements for reaction rate monitoring.

### 9.3.3. All-Electronic Reciprocal Time Computers

A hybrid system for use with variable-time applications was proposed by Crouch (86). Figure 9.27 shows the corresponding block diagram for this system. The system uses all solid-state electronics for rapid response and eliminates critical logarithmic elements. The system computes the reciprocal of the time by measuring the period of a pulse train whose frequency is directly proportional to the desired time interval. It requires a voltage input and operates with either positive- or negative-going signals.

A digital system for computing reciprocal time and with direct readout in either concentration or reciprocal time has been described by Pardue et al. (87). The digital circuitry generates a frequency proportional to reciprocal time and eliminates the need for analog time base and V-to-F conversion.

### 9.3.4. Systems for Differential Rate Measurements

Pinkel and Mark (88) proposed an analog system for automatic collection of data obtained using spectrophotometric monitoring and the method of proportional equations. The photocurrent generated at the photometric detector is converted by a logarithmic circuit to a voltage proportional to concentration, which in turn is fed into the memory and computing circuit (Figure 9.28). Two sample-and-hold circuits act as memory units. Time is the discriminating variable; at a certain time a programmed timer applies the signal held in memory to an analog circuit capable of solving simultaneous equations. The unit was tested with simulated rate curves but actual application to chemical determination was not reported.

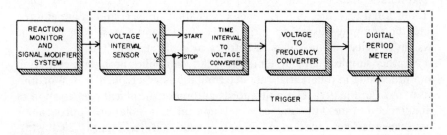

**Figure 9.27.** Block diagram of hybrid variable-time rate measurement system. [Reproduced with permission from S. R. Crouch, *Anal. Chem.*, **41**, 880. Copyright 1969 by the American Chemical Society.]

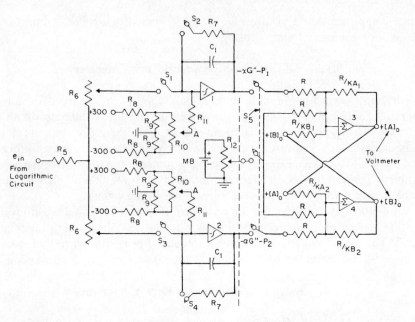

**Figure 9.28.** Memory and computing circuit for automatic treatment of spectrophotometric data by the method of proportional equations. Operational amplifiers 1 and 2 are integrators for sample and hold; a programmed timer sequences all events. Before preselected times $t_1$ and $t_2$ all switches are open with the exception of switches $S_2$ and $S_4$. At time $t_1$ relay $S_2$ opens and remains opened. Simultaneously $S_1$ closes for a time short in comparison with the overall reaction time. The signal at $t_1$ is thus connected to the sample-and-hold amplifier and held after switch $S_1$ reopens. Similarly, at $t_2$ the signal is sampled and held at the output of integrator 2. At a preset time after $t_2$, relay $S_5$ allows the output from both integrators to be entered into the computing unit (operational amplifiers 3 and 4). [From Ref. 88, reproduced with permission of Pergamon Press.]

A similar analog circuit was reported by Toren and Davis (89). Data proportional to values of a parameter (signal) at times $t_1$ and $t_2$ are entered into the computer using potentiometers. Toren and Gnuse (90) used the system to determine mixtures of ketones by a pH-stat method.

A rather simple analog system that greatly simplifies both data collection and processing in differential rate determinations was reported by Worthington and Pardue (91). The unit affords automatic graphical presentation of simultaneous rate data and uses a graphical extrapolation approach for simultaneous determination in two-component mixtures. Figure 9.29 illustrates the circuit for generating exponential functions; in connection with a time base it continuously generates variable values of $\exp(-k_A t)$ and $\exp(-k_B t)$. These functions drive the $x$ axis of an $x$–$y$ recorder, and a circuit providing a

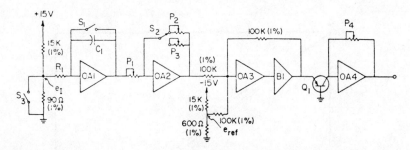

**Figure 9.29.** Circuit for generating the exponential functions used to drive the $x$ axis of an $x-y$ recorder in the analog system for application of graphical extrapolation in the analysis of binary mixtures by differential reaction rates. [J. B. Worthington and H. L. Pardue, *Anal. Chem.*, **44**, 767. Copyright 1972 by the American Chemical Society.]

voltage level proportional to concentration drives the $y$ axis. Figure 9.30 illustrates the operation of the system. With the $\exp(-k_A t)$ function driving the $x$ axis, the system is started when reactants are mixed. At the point when the plot starts to deviate from linearity the system is switched and the $\exp(-k_B t)$ function drives the $x$ axis. The system is allowed to continue to operate in this fashion until sufficient data are collected to permit reliable extrapolation of response of the slower-reacting component.

### 9.3.5. Analog Systems for Catalytic End-Point Detection and Miscellaneous Switching Systems

A very convenient switching circuit for automatic recording of the time elapsed to reach the "end point" in catalytic end point indication is illustrated in Figure 9.31. The circuit is wired to stop a timer in a titration in which a positive incoming signal decreases in absolute value after the end point, corresponding to a decrease in signal (concentration) due to the monitored species. If the situation is reversed (i.e., signal is positive and increasing in absolute value) all diodes need to be reversed. In the actual titration a buret with constant-rate delivery and the timer are automatically and simultaneously switched on. The signal level remains constant or changes very slowly, depending on the rate of the uncatalyzed reaction, until the rate of the indicator reaction is dramatically affected by the increasing (excess) amount of catalytic titrant. Consider the curve that plots concentration of monitored species versus time. If a pre-established signal value located in the catalytic portion of the curve is chosen as a reference potential in the switching network, so that the operational amplifier is driven to saturation when the signal is smaller in absolute value and of opposite sign to $E_{ref}$, the timer will

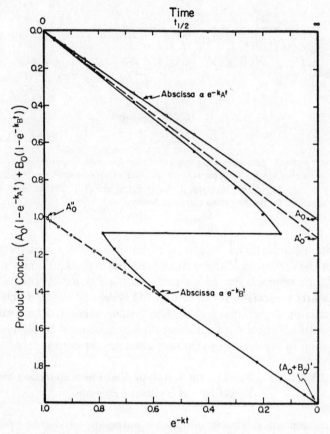

**Figure 9.30.** Product concentration as a function of exponential functions generated by the circuit of Figure 9.29. $k_a = 4 \times 10^{-2}\,\text{s}^{-1}$; $k_b = 5 \times 10^{-3}\,\text{s}^{-1}$; $\bigcirc$: $A_0 = 1$, $B_0 = 0$; $\bullet$, $\blacksquare$: $A_0 = B_0 = 1$. $\square$: $A_0 = 0$, $B_0 = 1$; [Reprinted with permission from J. B. Worthington and H. L. Pardue, *Anal. Chem.*, **44**, 767. Copyright 1972 by the American Chemical Society.]

be stopped and the time recorded will be proportional to the amount of analyte in the sample (92).

The circuit in Figure 9.31 represents a single-switch network; a double-switch configuration (Figure 9.32) provides a simple and convenient way to obtain the time interval for pre-established concentrations (signals) in the variable-time procedure. Analytically the circuit receives signal changes proportional to concentration from any transducer whose output can be converted to a voltage drop with respect to ground within the limits of the two reference potentials, $E_U$ and $E_L$ (92, 93). A solid-state double-switching network of similar operation has been used by Nikolelis and Hadjiioannou

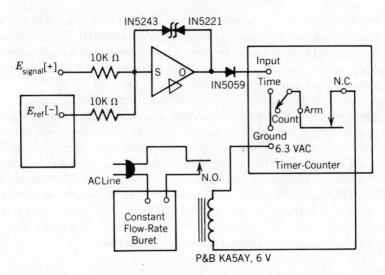

**Figure 9.31.** Single-switch circuit for automatic recording of time proportional to volume of titrant added in titrations with catalytic end-point indication. [From Ref. 92.]

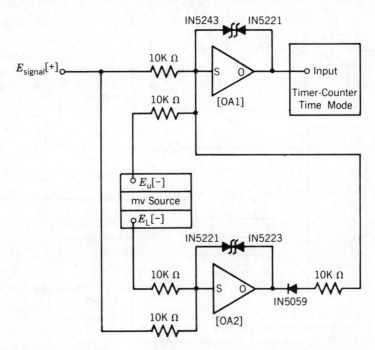

**Figure 9.32.** Double-switch network for implementing the variable-time approach. [From Ref. 92.]

(94) for the determination of inhibitors (amino polycarboxylic acids) of the manganese-catalyzed periodate oxidation of antimony(III). Details of their switching network are available (95, 96). A hybrid analog and digital double-switching electronic clock system, also for collection of variable-time data, has been reported by Chlapowski and Mottola (97).

The need for collecting information at different times of the reaction profile (e.g., different portions of a titration curve) may arise in the study of a given system and in the development of kinetic methods. The triple-switch network illustrated in Figure 9.33, for example, can be used to collect information during the inhibition portion of titrations with catalytic end-point indication as well as during the part in which catalysis is in operation (92). Operational amplifier OA1 acts as a simple polarity switch, and in connection with OA2 and the diodes between OA2 and OA3, prevents saturation of OA3 until $|E_{signal}| = |E_{ref1}|$. After $|E_{signal}| = |E_{ref1}|$, the network works exactly like the double-switch network described earlier in this section.

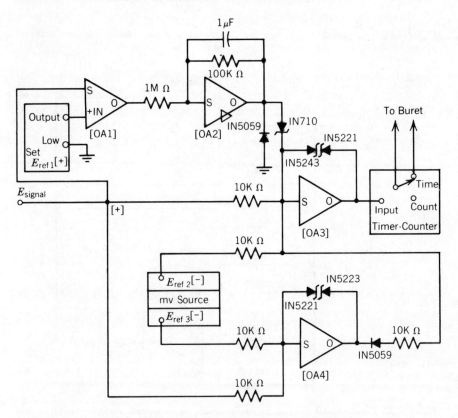

**Figure 9.33.** Triple-switch network. [From Ref. 92.]

## 9.4. COMPUTERS

A good introduction to digital electronics and microcomputers can be found in a chapter by Holler (98), and an in-depth discussion of electronics and microcomputer principles and practice has been given by Enke et al. (99). A very comprehensive monograph on computers in analytical chemistry is also available (100), and applications of small digital computers to kinetic-based analytical methods have been treated by Crouch (81).

An excellent discussion of the basics of computer experimentation for the study of complex chemical systems has been given by Frazer (101). This contribution has been followed up with a specific application of graphical techniques for the analysis of kinetic data of analytical interest (102).

Holtzman (103) has introduced a microprocessor-based microcomputer system specifically designed for rapid studies of fast reactions.

Computers dominate the third level of automation in the analytical laboratory (1) by making possible communication among units in the other two levels. Feedforward and feedback loops depend on computers for operation; computers also provide the sophistication needed for data manipulation and reduction.

Computers fulfill another important ancillary task in kinetic-based methods: the simulation of chemical reactions. An excellent account of the use of computers in the analysis and simulation of reactions has been given by Come (104). Edelson (105) has also reviewed numerical methods for modeling complex chemical reactions.

## REFERENCES

1. H. A. Mottola, *Anal. Chim. Acta*, **180**, 26 (1986); *Anal. Sci.*, **2**, 317 (1986).

2. R. G. McKee, *Anal. Chem.*, **42**, 91A (1970).

3. C. G. Enke, *Anal. Chem.*, **43**, 69A (1971).

4. H. V. Malmstadt, C. G. Enke, and E. C. Toren, Jr., *Experimental Electronics*, Prentice-Hall, New York, 1962.

5. K. B. Yatsimirskii, *Kinetic Methods of Analysis*, Pergamon, Oxford, 1966, pp. 29–30.

6. R. H. Conrad, *Anal. Chem.*, **39**, 1039 (1967).

7. T. J. N. Carter and B. R. Stanbridge, *Analyst (London)*, **103**, 968 (1978).

8. R. A. Harvey, *Anal. Biochem.*, **29**, 58 (1969).

9. P. Strittmatter, in *Rapid Mixing and Sampling Techniques in Biochemistry*, B. Chance, R. H. Eisenhard, Q. H. Gibson, and K. K. Lonberg-Holm, Eds., Academic Press, New York, 1964, p. 71.

10. F. J. Roughton and B. Chance, in *Investigation of Rates and Mechanisms of Reactions*, 2nd ed., Part II, S. L. Fries, E. S. Lewis, and A. Weissberger, Eds., Interscience, New York, 1963.

11. R. L. Berger, B. Balko, W. Borcherd, and W. Friauf, *Rev. Sci. Instrum.*, **39**, 486 (1968).

12. G. H. Czerlinski, *Chemical Relaxation*, Dekker, New York, 1966, p. 292.

13. S. R. Crouch, F. J. Holler, P. K. Notz, and P. M. Beckwitz, *Appl. Spectrosc. Rev.*, **13**, 165 (1977).

14. H. V. Malmstadt, K. M. Walczak, and M. A. Koupparis, *Am. Lab. (Fairfield, Conn.)*, **12** (9), 27 (1980).

15. M. A. Koupparis, K. M. Walczak, and H. V. Malmstadt, *J. Autom. Chem.*, **2**, 66 (1980).

16. F. J. Holler, C. G. Enke, and S. R. Crouch, *Anal. Chim. Acta*, **117**, 99 (1980).

17. J. E. Stewart, *Anal. Chem.*, **53**, 1125 (1981).

18. S. Stieg and T. A. Nieman, *Anal. Chem.*, **52**, 798 (1980).

19. I. R. Bonnell and J. D. Defreese, *Anal. Chem.*, **52**, 139 (1980).

20. E. F. Caldin, *Fast Reactions in Solution*, Wiley, London, 1964, Chapter 3.

21. H. A. Mottola, *Anal. Chem.*, **53**, 1312A (1981).

22. R. Stanley, *J. Autom. Chem.*, **6**, 175 (1984).

23. K. K. Stewart, *Talanta*, **28**, 789 (1981).

24. L. T. Skeggs, Jr., *Am. J. Pathol.*, **28**, 311 (1957).

25. L. Snyder, J. Levine, R. Stoy, and A. Conetta, *Anal. Chem.*, **48**, 942A (1976).

26. W. B. Furman, Ed., *Continuous-Flow Analysis: Theory and Practice*, Dekker, New York, 1976.

27. G. N. Stewart, *J. Physiol. (London)*, **22**, 159 (1897).

28. W. J. Blaedel and G. P. Hicks, *Anal. Chem.*, **34**, 388 (1962).

29. J. Ruzicka and E. H. Hansen, *Anal. Chim. Acta*, **78**, 145 (1975).

30. K. K. Stewart, G. R. Beecher, and P. E. Hare, *Anal. Biochem.*, **70**, 167 (1976).

31. J. Ruzicka and E. H. Hansen, *Flow Injection Analysis*, Wiley, New York, 1981.

32. M. D. Luque de Castro and M. Valcárcel, *Flow Injection Analysis: Principles and Applications*, Horwood, Chichester, U.K., 1987.

33. J. Slanina, W. A. Lingerak, and F. Bakker, *Anal. Chim. Acta*, **117**, 91 (1980).

34. H. A. Mottola and H. B. Mark, Jr., *Anal. Chem.*, **56**, 96R (1984).

35. H. A. Mottola, *Analyst (London)*, **112**, 717 (1987).

36. J. M. Reijn, W. E. van der Linden, and H. Pope, *Anal. Chim. Acta*, **126**, 1 (1981).

37. D. Betteridge, *Anal. Chem.*, **50**, 832A (1978).

38. C. C. Painton and H. A. Mottola, *Anal. Chim. Acta*, **154**, 1 (1983).

39. G. Horvai and E. Pungor, *CRC Crit. Rev. Anal. Chem.*, **17**, 231 (1987).

40. C. C. Painton and H. A. Mottola, *Anal. Chim. Acta*, **158**, 67 (1984).

41. C. C. Painton and H. A. Mottola, *Anal. Chem.*, **53**, 1713 (1981).

42. J. Ruzicka and E. H. Hansen, *Anal. Chim. Acta*, **161**, 1 (1984).

43. W. E. van der Linden, *Anal. Chim. Acta*, **180**, 20 (1986).

44. H. A. Mottola, *Anal. Chem.*, **42**, 630 (1970).

45. H. Hall, B. E. Simpson, and H. A. Mottola, *Anal. Biochem.*, **45**, 453 (1972).

46. C. F. Jacobsen, J. Leonis, K. Linderstrom-Lang, and M. Ottesen, in *Methods of Biochemical Analysis*, Vol. 4, D. Glick, Ed., Interscience, New York, 1957, p. 171.

47. H. Weisz and G. Fritz, *Anal. Chim. Acta*, **123**, 239 (1981).

48. S. Pantel and H. Weisz, *Anal. Chim. Acta*, **70**, 391 (1974); **89**, 47 (1977).

49. S. Pantel and H. Weisz, *Anal. Chim. Acta*, **74**, 275 (1975).

50. G. D. Nichols, *Anal. Chem.*, **53**, 489A (1981).

51. R. C. Coleman, W. D. Schultz, M. T. Kelly, and J. A. Dean, *Am. Lab. (Fairfield, Conn.)*, **3**(7), 26 (1971).

52. N. G. Anderson, *Fresenius' Z. Anal. Chem.*, **261**, 257 (1972).

53. C. D. Scott and C. A. Burtis, *Anal. Chem.*, **45**, 327A (1973).

54. T. E. Weichselbaum, W. H. Plumpe, Jr., and H. B. Mark, Jr., *Anal. Chem.*, **41** (3), 103A (1969).

55. T. E. Weichselbaum, W. H. Plumpe, Jr., R. E. Adams, J. C. Hagerty, and H. B. Mark, Jr., *Anal. Chem.*, **41**, 725 (1969).

56. H. L. Pardue and P. A. Rodriguez, *Anal. Chem.*, **39**, 901 (1967).

57. H. A. Mottola and H. B. Mark, Jr., "Kinetic Methods", in *Instrumental Analysis*, 2nd ed., G. D. Christian and J. E. O'Reilly, Eds., Allyn and Bacon, Boston, 1986, pp. 580–585.

58. G. Horlick, *Appl. Spectrosc.*, **30**, 113 (1976).

59. J. Cahill and M. Retzik, *Am. Lab. (Fairfield, Conn.)*, **16** (11), 47 (1984).

60. M. Valcárcel, *Analyst (London)*, **112**, 729 (1987).

61. J. D. Ingle, Jr., and M. A. Ryan, in *Modern Fluorescence Spectroscopy*, Vol. 3, E. L. Wehry, Ed., Plenum, New York, 1981, pp. 99–102.

62. M. Valcárcel and F. Grases, *Talanta*, **30**, 139 (1983).

63. W. R. Seitz, *CRC Crit. Rev. Anal. Chem.*, **13**, 1 (1981).

64. L. J. Kricka and G. H. G. Thorpe, *Analyst (London)*, **108**, 1274 (1983).

65. R. L. Wilson and J. D. Ingle, Jr., *Anal. Chem.*, **49**, 1060 (1977).

66. R. L. Wilson and J. D. Ingle, Jr., *Anal. Chim. Acta*, **83**, 203 (1976).

67. P. E. Stanley, in *Clinical and Biochemical Luminescence*, L. J. Kricka and T. J. N. Carter, Eds., Dekker, New York, 1982, p. 219.

68. E. L. Wehry, *Anal. Chem.*, **52**, 75R (1980); **54**, 131R (1982); **56**, 156R (1984); **58**, 13R (1986).

69. Y. Hayashi and H. Yuki, *Chem. Pharm. Bull.*, **32**, 3079 (1984).

70. Y. Hayashi, M. Ikeda, and H. Yuki, *Anal. Chim. Acta*, **167**, 81 (1985).

71. P. T. Kissinger, in *Laboratory Techniques in Electroanalytical Chemistry*, P. T. Kissinger and W. R. Heineman, Eds., Dekker, New York, 1984, Chapter 22.

72. H. A. Mottola, Ch. -M. Wolff, A. Iob, and R. Gnanasekaran, in *Modern Trends in Analytical Chemistry*, E. Pungor and I. Buzas, Eds., Analytical Chemistry Symposia Series, Elsevier, Amsterdam, 1984, pp. 49–75.

73. J. B. Sand and C. O. Huber, *Anal. Chem.*, **42**, 238 (1970).

74. K. J. Caserta, F. J. Holler, S. R. Crouch, and C. G. Enke, *Anal. Chem.*, **50**, 1534 (1978).

75. J. Jordan, J. K. Grime, D. M. Waugh, C. D. Miller, H. M. Cullis, and D. Lohr, *Anal. Chem.*, **48**, 427A (1976).

76. E. J. Greenhow, *Chem. Rev.*, **77**, 835 (1977).

77. S. A. Jacobs, G. W. Kramer, R. E. Santini, and D. W. Margerum, *Anal. Chim. Acta*, **157**, 117 (1984).

78. B. F. Simpson, J. R. Kincaid, F. J. Holler, *Anal. Chem.*, **55**, 1422 (1983).

79. H. V. Malmstadt, C. J. Delaney, and E. A. Cordos, *CRC Crit. Rev. Anal. Chem.*, **2**, 559 (1972).

80. H V. Malmstadt, C. J. Delaney, and E. A. Cordos, *Anal. Chem.*, **44** (12), 79A (1972).

81. S. R. Crouch, in *Computers in Chemistry and Instrumentation*, Vol. 3: *Spectroscopy and Kinetics*, J. S. Mattson, H. B. Mark, Jr., and H. C. MacDonald, Jr., Eds., Dekker, New York, 1973, Chapter 3.

82. H. L. Pardue, *Anal. Chem.*, **36**, 633 (1964).

83. H. V. Malmstadt and S. R. Crouch, *J. Chem. Educ.*, **43**, 340 (1966).

84. E. A. Cordos, S. R. Crouch, and H. V. Malmstadt, *Anal. Chem.*, **40**, 1812 (1968).

85. J. D. Ingle, Jr., and S. R. Crouch, *Anal. Chem.*, **42**, 1055 (1970).

86. S. R. Crouch, *Anal. Chem.*, **41**, 880 (1969).

87. H. L. Pardue, R. A. Parker, and B. G. Willis, *Anal. Chem.*, **42**, 56 (1970).

88. D. Pinkel and H. B. Mark, Jr., *Talanta*, **12**, 491 (1965).

89. E. C. Toren, Jr., and J. E. Davis, *Anal. Lett.*, **1**, 289 (1968).

90. E. C. Toren, Jr. and M. K. Gnuse, *Anal. Lett.*, **1**, 295 (1968).

91. J. B. Worthington and H. L. Pardue, *Anal. Chem.*, **44**, 767 (1972).

92. H. A. Mottola, *MPI Appl. Notes*, **6**, 17 (1971).

93. H. A. Mottola and G. L. Heath, *Anal. Chem.*, **44**, 2322 (1972).

94. D. P. Nikolelis and T. P. Hadjiioannou, *Anal. Chem.*, **50**, 205 (1978).

95. D. P. Nikolelis and T. P. Hadjiioannou, *Mikrochim. Acta*, **9**, 125 (1977).

96. C. E. Efstathiou and T. P. Hadjiioannou, *Anal. Chim. Acta*, **89**, 55 (1977).

97. E. W. Chlapowski and H. A. Mottola, *Anal. Chim. Acta*, **76**, 319 (1975).

98. F. James Holler, "Digital Electronics, Data-Domain Conversions, and Microcomputers," in *Instrumental Analysis*, 2nd ed., G. D. Christian and J. E. O'Reilly, Eds., Allyn and Bacon, 1986, Boston, Chapter 24.

99. C. G. Enke, S. R. Crouch, F. J. Holler, H. V. Malmstadt, and J. P. Avery, *Anal. Chem.*, **54**, 367A (1982).

100. P. Barker, *Computers in Analytical Chemistry*, Pergamon, New York, 1983.

101. J. W. Frazer, *Anal. Chem.*, **52,** 1205A (1980).

102. J. W. Frazer and H. R. Brand, *Anal. Chem.*, **52,** 1730 (1980).

103. J. L. Holtzman, *Anal. Chem.*, **52,** 989 (1980).

104. G. M. Come, in *Comprehensive Chemical Kinetics*, Vol. 24, C. H. Bamford and C. F. G. Tipper, Eds., Elsevier, Amsterdam, 1983, Chapter 3.

105. D. Edelson, *Science*, **214,** 981 (1981).

# CHAPTER

## 10

# ERROR ANALYSIS IN KINETIC-BASED DETERMINATIONS

The number of published papers that consider the limitations associated with kinetic measurements on a mathematical basis is relatively limited. However, the importance of the subject deserves a separate chapter. The elegance of mathematical treatment, however, should not obscure the fact that chemical reaction rate methods, for effective analytical use and full exploitation of their analytical potential, still require an understanding of the chemistry involved, particularly reaction mechanisms. Such knowledge also helps in choosing optimal conditions for analytical determination.

## 10.1. MINIMIZATION OF SYSTEMATIC AND RANDOM FLUCTUATIONS IN RATE COEFFICIENTS

### 10.1.1. Single-Rate Measurement Approach

As mentioned in the bulk of the material covered in preceding chapters, most kinetic-based determinations are performed under first-order or pseudo-first-order conditions. Hewitt and Pardue (1) pointed out that under such conditions the rate coefficient becomes a unique diagnostic device to evaluate sample-to-sample variations. This is because such a coefficient is independent of the concentration of the rate-limiting species, as can be simply shown. Thus uncertainties regarding the rate coefficient can be considered an important source of errors in kinetic methods based on first-order or pseudo-first-order reaction conditions. The relative uncertainty in the rate coefficient resulting from fluctuations in pH, for instance, can amount to almost 5% if the uncertainty in the pH is 0.02 pH units (2).

Assume the simple reaction

$$A + B \rightarrow P$$

If the analyte A is rate-limiting and B is in sufficiently large excess, the concentration–time profile expected will be that shown in Figure 10.1 (the

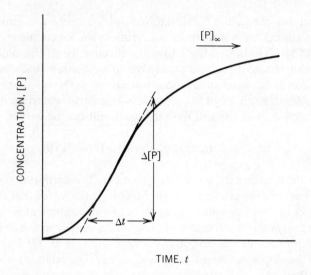

**Figure 10.1.** Concentration–time profile for the reaction $A + B \rightarrow P$ under pseudo-first-order conditions.

monitored signal is directly proportional to [P]); thus the pseudo-first-order behavior can be expressed as

$$\text{rate} = \frac{d[P]}{dt} = k[P] = -k[A] = k[A]_0 e^{-kt} \tag{10.1}$$

where $k$ = first-order rate coefficient. In the vicinity of the inflection point (see Figure 10.1) the response is approximately linear, implying an apparent zero-order behavior, and

$$[A]_0 \simeq -(1/k)(\Delta[P]/\Delta t) \tag{10.2}$$

Integrating eq. 10.1 we have

$$\ln([P]_t/[P]_0) = kt \tag{10.3}$$

and

$$k = \Delta[\ln([P]_\infty - [P]_t)]/\Delta t \tag{10.4}$$

Plots of $\ln([P]_\infty - [P]_t)$ versus $t$ should be linear and the slope equal to the rate coefficient $k$, independent of $[A]_0$. Hence, if values of $k$ determined with different samples containing different concentrations of analyte within certain limits, are for all practical purposes constant, we can conclude that expected small variations in reaction conditions (e.g., pH, temperature, ionic strength, matrix variations) are not introducing errors in the kinetic determination.

Landis et al. (3) indicated that "to make effective use of this built-in 'error diagnostic,' it is useful to have an analytical expression that describes the concentration in terms of uncertainties or errors in the rate constant." The relative concentration error at any time $t$ (for the determination of A) resulting from uncertainties in the rate coefficient can be written as

$$\text{relative concentration error} = [t - (1/k)]\sigma_k \qquad (10.5)$$

where $\sigma_k$ = the standard deviation of experimentally determined values of $k$; it also represents the uncertainty in $k$ (3). Equation 10.5 shows that the effect of uncertainties in $k$ values is minimized if the rate measurement is made at the time when $t = 1/k$. Since a single-point measurement is expected to suffer from random variations in the response curve, Landis et al. (3) recommended averaging the slopes measured at several points. The relative concentration error will then be

$$\text{relative concentration error} = \left[\sum_i^n t_i - (n/k)\right](\sigma_k/n) \qquad (10.6)$$

where $n$ = total number of points averaged. Using the theory of propagation of errors, Holler et al. (4) confirmed that at $t = 1/k$, reaction rates are essentially independent of small fluctuations in parameters affecting the value of $k$. The propagation-of-errors treatment allows one to ascertain the effect of random errors starting with the expression for the variance of a given measured rate at any time $t$:

$$\sigma^2_{(\text{rate})_t} = (\partial_{(\text{rate})_t}/\partial k)^2 \sigma_k^2 \qquad (10.7)$$

From eq. 10.1 we can write

$$\partial_{(\text{rate})_t}/\partial k = [A]_0 e^{-kt}(1 - kt) \qquad (10.8)$$

thus

$$\sigma^2_{(\text{rate})_t} = [A]_0^2 e^{-2kt}(1 - kt)^2 \sigma_k^2 \qquad (10.9)$$

and

$$(\sigma_{(\text{rate})_t}/\text{rate}) = |1 - kt|(\sigma_k/k) \qquad (10.10)$$

Equation 10.10 shows that the relative imprecision in the measurement of the rate at time $t$ is zero at $t = 1/k$. Experimental results obtained with the reaction between iodide ion and hydrogen peroxide agree very well with the theoretical forecast. Similarly, systematic errors in $k$ lead to

$$\Delta(\text{rate})/\text{rate} = (1 - kt)(\Delta k/k) \tag{10.11}$$

indicating that the effects of systematic errors in the measurement of $k$ are also eliminated or minimized by rate measurements at $t = 1/k$. All these considerations, however, assume (a) that the rate is measured at $t = 1/k$, which may not be possible in certain chemical systems; (b) that there is zero error in the measurement of time (a not unrealistic assumption with timing devices available today), and (c) that there is negligible instrumental imprecision.

Hence it is possible to estimate $\sigma_k$ and $\Delta k$ by applying the theory of propagation of errors and noting the dependence of $k$ on experimental parameters affecting it, $\rho_i$, (e.g., temperature, ionic strength, pH) from

$$\sigma_k^2 = \sum_{i=1}^{n} (\partial k/\partial \rho_i)^2 \sigma_{\rho_i}^2 \tag{10.12}$$

and

$$\Delta k = \sum_{i=1}^{n} (\partial k/\partial \rho_i) \Delta \rho_i \tag{10.13}$$

The foregoing considerations imply that in methods based on first-order or pseudo-first-order reactions a fixed-time procedure with measurements at time $t = 1/k$ should be superior to, for instance, initial rate measurements. The initial rate of the pseudo-first-order reaction presented at the beginning of this section can be expressed as

$$\text{rate}_{t=0} = k_T [\text{A}]_0 \tag{10.14}$$

Similarly at $t = 1/k_{\bar{T}}$, where $T$ is the experiment temperature at the hypothetical time $t = 0$ and $\bar{T}$ is the average temperature during the time span of the experiment, this pseudo-first-order rate becomes

$$\text{rate}_{t=(1/k),\bar{T}} = k_T [\text{A}]_0 \exp(-k_T/k_{\bar{T}}) \tag{10.15}$$

Simulated plots of these rates as a function of $[\text{A}]_0$ are shown in Figure 10.2. The plot assumes a linear, systematic change of temperature (systematic error effect) during the course of the experiment from $\Delta T$ below $\bar{T}$ to $\Delta T$ above $\bar{T}$. Values of $\Delta T$ of $\pm 0.5$ have no noticeable effect on the rates measured at $t = 1/k$, but initial rates show a steadily increasing slope as the temperature interval increases. The effect of normal random thermal noise on the same rates is shown in the simulated curves of Figure 10.3. The differing thermal noise fluctuations are indicated in the legend of the figure. Again,

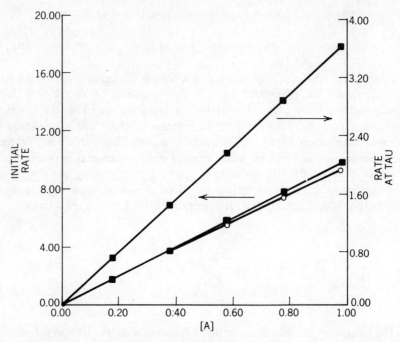

**Figure 10.2.** Simulated data comparing calibration curves generated with rates measured at $t = 1/k$ (rate at TAU) and with curves generated at $t = 0$ (initial rate). Systematic errors in temperature linearly increased with time from $300\ K - \Delta T$ to $300\ K + \Delta T$. Values of $\Delta T$ used: 0.1 K, 0.2 K, 0.3 K, 0.4 K, and 0.5 K. [Reprinted with permission from F. J. Holler et al., *Anal. Chem.*, **54**, 755. Copyright 1982 by the American Chemical Society.]

temperature fluctuations have no apparent effect on rate measurements at $t = 1/k$ but a clear effect on measurements at $t = 0$. Similar effects can be expected for systematic and random fluctuations of any experimental parameter affecting $k$. One of these factors is concentration of reactants and catalyst (when pertinent). Atwood and DiCesare (5) pointed out, for example, that the effect of rate coefficient fluctuations can be minimized by adjusting the enzyme activity in enzymatic substrate determinations so that the reciprocal of the pseudo-first-order rate coefficient equals the time of measurement.

### 10.1.2. Two-Rate Measurement Approach

All foregoing considerations imply that to minimize kinetic fluctuations, a fixed-time measurement at (or as close to) $t = k$ should suffice. Rate measurements at two different times, however, make the method independent of

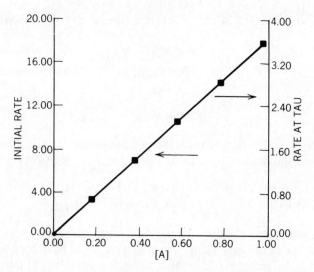

**Figure 10.3.** Simulated data comparing calibration curves derived from rates measured at $t = 1/k$ (rate at TAU) with curves generated from rates measured at $t = 0$ (initial rate). Values of random temperature fluctuations $\sigma_T$ used: 0.1 K, 0.2 K 0.3 K, 0.4 K, and 0.5 K. [Reprinted with permission from F. J. Holler et al., *Anal. Chem.*, **54**, 755. Copyright 1982 by the American Chemical Society.]

changes in rate coefficient (6). Such an independence can be mathematically justified by realizing that application of eq. 10.1 to rate measurements at two times $t_1$ and $t_2$ during the course of the reaction leads to

$$\text{rate}_{t_1}/\text{rate}_{t_2} = e^{k\Delta t} \tag{10.16}$$

such that

$$k = [\ln(\text{rate}_{t_1}/\text{rate}_{t_2})]/\Delta t \tag{10.17}$$

Substituting eq. 10.16 into eq. 10.1 and rearranging gives an expression for the initial concentration of the sought-for species that is independent of the value of the rate coefficient:

$$[A]_0 = \frac{\text{rate}_{t_1} \Delta t}{\ln(\text{rate}_{t_1}/\text{rate}_{t_2})} \left(\frac{\text{rate}_{t_1}}{\text{rate}_{t_2}}\right)^{t_1/\Delta t} \tag{10.18}$$

Derivation of this equation is made, of course, with the assumption that $k$ is the same at times $t_1$ and $t_2$. Since two rates instead of one are measured and a nonlinear function needs to be evaluated, it is important to consider the effect of random errors. As errors in time measurement can be considered negligible, the variance in the two-rate parameter can be written as

$$(\sigma[A]_0)^2 = \left\{ \frac{\partial[A]_0}{\partial(rate)_{t_1}} \right\}^2_{(rate)_{t_2}} [\sigma_{(rate)_{t_2}}]^2$$
$$+ \left\{ \frac{\partial[A]_0}{\partial(rate)_{t_2}} \right\}^2_{(rate)_{t_1}} [\sigma_{(rate)_{t_1}}]^2 \qquad (10.19)$$

Assuming equal variances in all rate measurements (a valid assumption in many cases), the relative variance in the measured concentration is given by

$$\frac{\sigma^2[A]_0}{[A]_0^2} = \frac{e^{2t_1}[(t_1 + \Delta t - 1)^2 + e^{2\Delta t}(1 - t_1)^2]}{e^2(\Delta t)^2} \cdot \frac{[\sigma_{(rate)_{t = 1/k}}]^2}{[(rate)_{t = 1/k}]^2} \qquad (10.20)$$

which is a function of the times $t_1$ and $t_2$ at which the rate measurements are performed. All of the time measurements in eq. 10.20 have been adjusted to units of $1/k$ to remove the explicit dependence of the equation on the magnitude of the rate coefficient. Equation 10.20 can be simplified for limiting cases; Table 10.1 lists the functional forms for the simplified cases, the formula for the relative variances, and the times suggested for measurements. As could be expected, results are less precise when the time interval between measurements is short or when measurements are made at extended time intervals. Wentzell and Crouch (6) tested the two-rate measurement method with two chemical systems: (a) the effect of pH variations on the metal–complex exchange reaction involving 1,2-cyclohexanediamine-$N,N,N',N'$-tetraacetate as ligand (7), and (b) the formation of 12-molybdophosphate used in phosphate determinations (8) to verify the effect of thermal variations. The general trends of experimental results were in good quantitative agreement with the theoretical predictions.

### 10.1.3. Multirate Measurements: Multipoint Linear Regression Treatment with Predictive Equilibrium Signal Values

A different approach to minimizing errors introduced by changes in rate coefficient due to variables such as pH and temperature, under first- or pseudo-first-order conditions, was proposed by Mieling and Pardue (9). A multipoint linear regression program is used to calculate the values of rate coefficient, initial concentration, and concentration at equilibrium that fit experimental data to a first-order model. The analyte concentration is computed from the difference between initial and final (equilibrium) signal values computed from the best fit:

$$\Delta \hat{S} = \hat{S}_\infty - \hat{S}_0 \qquad (10.21)$$

**Table 10.1. Functional Form for Calculating Analyte Initial Concentration by Two-Rate Method**

| Limiting Case | Functional Form | Formula for Relative Variance[a] | Suggested Measurement Times ($\tau = 1/k$) |
|---|---|---|---|
| Simplified: $t_1 = 0,\ t_2 = t$ | $[A]_0 = \dfrac{R_0 t}{\ln(R_0/R_t)}$ | $\dfrac{(t-1)^2 + e^{2t}}{e^2 t^2} b^2$ | $t = 0.8\tau$ to $1.5\tau$ |
| Extended: $t_1,\ t_2 = 2t_1$ | $[A]_0 = \dfrac{R_{t_1}^2 \Delta t}{R_{t_2} \ln(R_{t_1}/R_{t_2})}$ | $\dfrac{e^{2t_1}(t_2 - 1)^2 + e^{4t_1}(1 - t_1)^2}{e^2 t_1^2} b^2$ | $t_1 = \Delta t = 0.5\tau$ to $0.8\tau$; $t_2 = 2t_1$ |
| Generalized: $t_1,\ t_2$ | $[A]_0 = \dfrac{R_{t_1} \Delta t}{\ln(R_{t_1}/R_{t_2})} \left(\dfrac{R_{t_1}}{R_{t_2}}\right)^{t_1/\Delta t}$ | $\dfrac{e^{2t_1}[(t_2 - 1)^2 + e^{2\Delta t}(1 - t_1)^2]}{e^2 (\Delta t)^2} b^2$ | $t_1 = 0.4\tau$ to $0.9\tau$; $\Delta t = 0.5\tau$ to $1.8\tau$ |

[a] $b = \sigma(R_t)/R_t = \mathrm{RSD}(R_t)$. All times are in units of $\tau = 1/k$.

Source: Adapted from P. D. Wentzell and S. R. Crouch, *Anal. Chem.*, **58**, 2851. Copyright 1986 by the American Chemical Society.

assuming $S$ to be linearly related to analyte concentration. Figure 10.4 illustrates the conceptual basis of the approach. The best fit is extracted by minimization of the following function:

$$\chi^2 = \frac{1}{s_S^2} \sum \left( S_t - S_t^\circ - \frac{\partial S_t^\circ}{\partial k} \partial k - \frac{\partial S_t^\circ}{\partial S_0} \partial S_0 - \frac{\partial S_t^\circ}{\partial S_\infty} \partial S_\infty \right)^2 \qquad (10.22)$$

where $S_t^\circ =$ an initial estimate of $S_t$ in terms of initial estimates of $S_0$, $S_\infty$, and $k$; and $s_S =$ the estimated standard deviation of signal measurements. No weighting function is needed if the uncertainty in $S$ is relatively constant throughout the range. The function is minimized by setting the first derivatives of $\chi^2$ with respect to $\Delta S_0$, $\Delta S_\infty$, and $\Delta k$ equal to zero and solving the resulting normal equations for parameters $\Delta S_0$, $\Delta S_\infty$, and $\Delta k$. First estimates of $\hat{S}_0$, $\hat{S}_\infty$, and $\hat{k}$ will normally be in error and are improved by successive approximations. Iterations are continued until the change in the value of $\chi^2$ is less than 0.05% between successive iterations. If the first estimated values are poor, the starting point will be outside the region for rapid convergence. This is avoided by using an algorithm (10, 11) that senses the condition and the method of steepest descent to approach the approximate parabolic $\chi^2$ hypersurface where the regression method takes over.

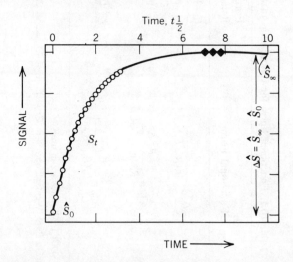

**Figure 10.4.** Illustration of multiple regression method. Open diamonds represent selected data points measured during part of the reaction. Solid diamonds represent near-equilibrium results used for comparison with kinetic results. Solid line is a fit of open diamond data to linearized first-order model with a multiple linear regression method (10). [Reprinted with permission from G. E. Mieling and H. L. Pardue *Anal. Chem.*, **50**, 1611. Copyright 1978 by the American Chemical Society.]

Several multipoint regression data-processing approaches have been evaluated by Pardue et al. (12–16). Particularly interesting is the use of multipoint nonlinear regression in substrate determination by enzyme-catalyzed reactions following the Michaelis–Menten equation. This is because the linear range of calibration curves can be extended beyond the value of the Michaelis–Menten constant $K_M$ (16). The data-processing approach in this case involves fitting data for absorbance (concentration) and rate versus time to the rate form of the Michaelis–Menten equation. The fitting parameters are the maximum initial rate, $K_M$, and the absorbance change from initial to equilibrium values that would be observed if the reaction were followed to completion. By assuming absorbance monitoring and a chemical system obeying the Michaelis–Menten scheme, it can be shown that the rate of absorbance change is

$$\frac{dA_t}{dt} = \frac{(IR)_{max}(\Delta A_\infty - \Delta A_t)}{K_M + (1/\varepsilon b)(\Delta A_\infty - \Delta A_t)} \tag{10.23}$$

where $dA_t/dt =$ the instantaneous rate of absorbance change at any time $t$, $\Delta A_t =$ the difference between absorbances at times $t$ and $t = 0$, $\Delta A_\infty =$ the absorbance change between $t = 0$ and $t = \infty$, $\varepsilon =$ the molar absorptivity of monitored species, $b =$ the optical path length, and $(IR)_{max} =$ the maximum initial rate. The method of Savitzky and Golay (17) is used to compute $\Delta A_t$ and $\Delta A_t/dt$ from experimentally collected multiple absorbance data plotted against time. The values of $\Delta A_t$ and $dA_t/dt$ are then fitted to eq. 10.23 with the multiple regression method (9) using $(IR)_{max}$, $K_M$, and $\Delta A_\infty$ as fitting parameters. Computed parameters of $\Delta \hat{A}_\infty$ should be proportional to initial substrate concentration. Hamilton and Pardue's report that for data corresponding to 70% or more completion for the most concentrated sample, calibration curves are linear for substrate concentrations up to 3.5 times the value of $K_M$ (16). A short review of kinetic methods based on predicted equilibrium values has been given by Harris and Hultman (18). They point out that, the first method proposed for calculating $A_\infty$ from data collected during the early part of the reaction dates back to 1926 and was proposed by Christie Smith (19). The method required measurement of absorbance at three different times.

The linear range of substrate determination is commonly considered to be limited to substrate concentrations that are small in comparison to the Michaelis–Menten constant by assuming a hyperbolic rate–concentration profile. Goren and Davis (20) have shown, however, that the rate–concentration curve for noninitial measurements is sigmoidal and that linear data can be obtained for substrate concentrations greater than $K_M$ by adequate selection of the enzymatic activity of the reagent and the measurement interval.

Czervionke et al. (21) have investigated the application of multipoint kinetic curve-fitting methods in the determination of creatinine by the Jaffé method and in the detection of acetoacetate interference in such determination.

### 10.1.4.  Optimization of Irreversible, Coupled First-Order Reactions

A rather common practice in enzymatic reactions (as indicated in Chapter 3) is the coupling of two enzymatic reactions. The kinetics of these coupled reactions have been studied by Bergmeyer (22), and different aspects of their optimization by minimization of errors have been addressed by Davis and Pevnick (23). Such consecutive, irreversible processes are also involved in other analytical techniques such as nonflame atomic absorption and continuous-flow analyses (24). (See Chapter 11.)
If

$$A \xrightarrow{\ k_1\ } B \xrightarrow{\ k_2\ } C \qquad (10.24)$$

the rate of final product formation is given by

$$\frac{d[C]}{dt} = [A]_0 k_1 \exp(-k_1 t) \left[ \frac{k_2}{k_2 - k_1} \right] \{1 - \exp[(-k_2 - k_1)t]\} \quad (10.25)$$

It should be noted that initial rate measurements have no meaning in this case because the initial rate is zero; there is a lag period during which the intermediate B builds up. An alternative is to select a region of the signal–time plot that approximates a straight line. In seeking to locate the "optimum" time when the greatest precision is obtained, Davis and Pevnick (23) set the derivative of the rate of production of C (eq. 10.25) equal to zero and solved the resulting equation for time. Since there are two variables, $k_1$ and $k_2$, the derivative can be taken with respect to one or the other or an infinite number of combinations of the two.

The derivative of the rate with respect to $k_1$ set to zero is

$$0 = 1 - k_1 t_1 - \frac{k_1}{k_2 - k_1} - k_1 t_1 \exp[-(k_2 - k_1)t_1]\{1 - \exp[-(k_2 - k_1)t_1]\}$$

$$(10.26)$$

where $t_1$ is the optimum time at which to measure the rate of the reaction. The optimization with respect to $k_1$ is valid for small changes in $K_M$, $(IR)_{max}$, or both, since

$$dk_1 = \left[ \frac{\partial k_1}{\partial(\text{rate})_{\text{max}}} \right] d(\text{rate})_{\text{max}} + \left( \frac{\partial k_1}{\partial K_\text{M}} \right) dK_\text{M} \qquad (10.27)$$

and

$$\frac{\partial k_1}{\partial(\text{rate})_{\text{max}}} = \frac{1}{K_\text{M}} \quad \text{as well as} \quad \frac{\partial k_1}{\partial K_\text{M}} = -\frac{(\text{rate})_{\text{max}}}{(K_\text{M})^2} \qquad (10.28)$$

since the derivatives differ only by a constant and would be zero for the same time. This also applies to the derivative with respect to $k_2$, which if set to zero gives

$$0 = \frac{-k_1}{k_2 - k_1}\{1 - \exp[-(k_2 - k_1)t_{\text{II}}] + k_2 t_{\text{II}} \exp[-(k_2 - k_1)t_{\text{II}}]\} \qquad (10.29)$$

where $t_{\text{II}} = $ optimum time at which to measure the rate of reaction.

Davis and Pevnick also consider the case of variations in both $k_1$ and $k_2$ starting with the derivative of the rae in its most general form:

$$d(\text{rate}) = (\partial \, \text{rate}/\partial k_1) \, dk_1 + (\partial(\text{rate})/\partial k_2) \, dk_2 \qquad (10.30)$$

For the particular case of dilution errors,

$$dk_2 = (k_2/k_1) \, dk_1$$

Substituting this into eq. 10.30, and setting $d(\text{rate}) = 0$ gives an equation that, as in the other cases, can be solved by Newton's method (25) to find the optimum time. Optimization with respect to rate coefficients implies optimization with respect to enzyme activity for each of the coupled reactions.

## 10.2. REGRESSION COMPUTATIONS IN DIFFERENTIAL REACTION RATE METHODS

Although the foregoing considerations on errors in first-order and pseudo-first-order reactions apply to differential rates determined under such conditions, some specific aspects applying to these methods have been chosen to be treated separately for didactic reasons. The second monograph published on the topic of kinetics in analytical chemistry (26) contains a chapter including general error analyses for the graphical extrapolation approach, the method of proportional equations, and first-order and pseudo-first-order methods when concentrations of reactants are much less than concentrations of reagents. Although intuitive thinking may lead the practicing chemist to approximately the same conclusions, reading the material covered in this chapter is recommended to the interested reader.

Ridder and Margerum (27) have emphasized the importance of using the entire signal–time response rather than using graphical extrapolation methods for the slowest-reacting component, initial rate measurements for only the fastest-reacting component, or two-point methods. The on-line computer processing of data from stopped-flow mixing and spectrophotometric determination of mixtures of metal ions, followed by a linear least-squares fit of the regression equation also done on-line (28) illustrates this point. Present-day availability of computers makes these tasks possible. Improved techniques for acquiring, calculating, and testing sets for simultaneous kinetic determinations have been suggested (27), and a set of recommendations is listed in Table 10.2. Early regression analysis computations using empirical weighting factors (28) have been improved by (a) a variable rate of data acquisition based on the value of component's reaction rate coefficient, (b) the technique of *centering the data* (29), and (c) a correlation matrix (29) used to obtain the

#### Table 10.2. General Recommendations for Differential Rate Methods

- Collection of multiple data rates should give approximately the same number of data points per three half-lives of each component.
- Weighting factors are not recommended unless the product of the rate coefficient for the fastest-reacting component multiplied by the uncertainty of the mixing time exceeds 0.1.
- Calibration runs should be performed with pure components and under exactly the same conditions as in determination runs.
- Linear least-squares regression analysis is possible for up to five components with on-line operation using minicomputers.
- The technique of centering the data and the use of a correlation matrix are recommended.
- Tests in which the residuals are displayed against the data points should be performed to detect incorrect or missing rate coefficients.
- The ratio of the concentrations of the components being determined should not greatly exceed the ratio of their rate coefficients.
- Nonlinear least squares can be used for a small number of components if the rate coefficients of the pure components under reaction conditions are not known accurately.
- Synergistic effects should be totally absent, and the kinetics of the system should be well defined.

*Source*: From Ref. 27.

best solution for all components in the mixture. Centering the data is used to make the absolute size of the numbers in the calculations smaller by considering only the time-dependent signal. This technique also reduces the order of the matrix by 1, eliminating constant background signals from the simultaneous equations.

The approach can be summarized as follows. With computer data handling one can easily treat a system where each reactant and each product are different and each has a measurable absorbance. The general situation is illustrated by eqs. 10.31:

$$A \xrightarrow{k_A} P_A$$

$$B \xrightarrow{k_B} P_B \tag{10.31}$$

$$C \xrightarrow{k_C} P_C$$

The total product of these concurrent reactions at any time $t_j$ is given by:

$$[P]_j = [A]_0\, a_j + [B]_0 b_j + [C]_0 c_j \tag{10.32}$$

where

$$a_j = [1 - \exp(-k_A t_j)]$$
$$b_j = [1 - \exp(-k_B t_j)]$$
$$c_j = [1 - \exp(-k_C t_j)] \tag{10.33}$$

Assuming photometric monitoring, the absorbance at any time is given by

$$\alpha_j = a_j \Delta A + b_j \Delta B + c_j \Delta C + \beta \tag{10.34}$$

where

$$\Delta A = [A]_0\,(\varepsilon_{P_A} - \varepsilon_A)l \qquad \Delta B = [B]_0\,(\varepsilon_{P_B} - \varepsilon_B)l$$
$$\Delta C = [C]_0\,(\varepsilon_{P_C} - \varepsilon_C)l \tag{10.35}$$

where $\varepsilon$ = the corresponding molar absorptivities, $l$ = the optical path length, and $\beta$ = a time-independent absorbance. It is possible to manipulate chemical systems in which one reaction causes an increase in absorbance and another reaction (with a different value of $k$) causes a decrease in absorbance. However, the kinetics must be known, and side reactions or synergistic effects, must be absent.

### 10.2.1. Least-Squares Regression

The data consists of $n$ pairs of observations, $\alpha_j, t_j$ ($j = 1, 2, \ldots, n$) taken over a given period of time. The residuals $r_j$ (difference between observed and expected absorbance values) are

$$r_j = \alpha_{j,\text{calc}} = \alpha_j - (\Delta A a_j + \Delta B b_j + \Delta C c_j + \beta) \qquad (10.36)$$

If the errors in measurement are random, the residuals will also be random and each residual will belong to a Gaussian distribution (standard deviation $\sigma_j$) centered around the true residual $\bar{r}_j = 0$.

The probability of finding a set of $n$ observations with unequal precision (unique $\sigma$ for each observation) is

$$\text{probability} = \left(\frac{dr_j}{\sqrt{2\pi}}\right)^n \frac{1}{\sigma_1 \sigma_2 \sigma_3 \cdots \sigma_n} \exp\left(-\frac{1}{2}\sum \frac{r_j^2}{\sigma_j^2}\right) \qquad (10.37)$$

The best fit of the regression equation is obtained at a minimum value for this probability. Minimization is obtained by adjusting $\Delta A$, $\Delta B$, $\Delta C$, and $\beta$ so that

$$\sum_{j=1}^{n} r_j^2 / \sigma_j^2$$

is minimum. If all standard deviations are equal, only minimization of $\Sigma r_j^2$ is required. In the case of unequal standard deviations the weighting factor $w_j = 1/\sigma_j^2$ is defined, and

$$\sum_{j=1}^{n} w_j r_j^2$$

should now be minimized. Ritter and Margerum (27) give details for the minimization of $\Sigma r_j^2$; they also illustrate the benefits of using the correlation matrix in defining the minimum of the error surface and the pitfalls of empirical weighting factors.

### 10.2.2. Nonlinear Regression

No linear relationship of the parameters can be established if one or more of the rate coefficients are included as adjustable parameters. In such a case no exact solution is possible and an iterative approach must be used. If the experimentalist is confronted with this situation, it is possible to (a) search the error surface for the minimum by a method of steepest descent, or (b) obtain the least-squares best estimate of the parameters using Taylor's expansion series to develop equations in which an exact solution for the estimate of the errors in the parameters can be obtained (30).

Ridder and Margerum (27) claim that excellent results can be obtained for a two-component system so long as the rate coefficients remain separated by a factor of 5 and a good signal-to-noise ratio (e.g., $S/N = 20$) is maintained.

Undoubtedly stopped-flow mixing is needed to permit collecting and processing as much rate data as possible for the fastest-reacting components in a mixture. The on-line computer processing of stopped-flow data has been described in detail by Willis et al. (31). Roos et al. (32) have described the interfacing of a relatively inexpensive microcomputer (Apple II) with a stopped-flow spectrophotometer for use in collecting kinetic data. They discussed use of this setup to collect and analyze multicomponent kinetic transients in simulated data and in the dissociation kinetics of the ethidium–DNA complex. The paper provides information on programs to collect and fit the kinetic data, on machine-language software, and on basic routines. The software-selectable sampling frequencies are from 0.1 ms to 8 s. Because of the large number of data points that can be collected and the ability to selectively obtain transmittance values in regions where the signal changes rapidly with time, relatively unsophisticated methods of data treatment can be used. Thompsen and Mottola (33) have described a microcomputer-based data acquisition system for stopped-flow mixing capable of acquiring signals with amplitudes of 50 mV to 3 V at rates of $5 \times 10^3$ to $10^{-1}$ points per second.

Interactive graphical computer techniques are useful for the display of response and residual surfaces as support for nonlinear techniques. Good examples of the utility of response surface methods have been provided by Frazer and Brand (34) and Burtis et al. (35).

## 10.3. COMPARATIVE ERROR STUDIES

Ingle and Crouch (36) have pondered the relative values of fixed-time and variable-time methods. Their theoretical treatment, experimentally verified, shows that the choice of rate measurement (fixed- or variable-time) depends on the characteristics of the reaction used, the kinetic role of the species under determination, and the relationship between signal and concentration of the transducer used for monitoring the reaction progress. The fixed-time approach is shown to be preferred for first-order or pseudo-first-order uncatalyzed reactions and for the determination of substrate concentration in enzyme-catalyzed reactions. Determinations of catalysts (including enzymes) are better served by the variable-time procedure, which is better suited for nonlinear response curves.

The signal-to-noise ratio theory for fixed- and variable-time procedures using photometric monitoring has been developed (37). The treatment is oriented to few-point data collection and was inspired by the ascertion that "one of the major limitations of spectrophotometric reaction rate measurements compared to equilibrium-based measurements is that the $S/N$ of the measurement is inherently smaller since only a portion öf the total available signal is utilized." Multipoint measurements (see above) have successfully removed such a limitation. Expressions for $S/N$ are useful, however, since they provide a systematic means for locating factors limiting precision. The most useful definition of $S/N$ for kinetic-based measurements is based on the following:

$$S = \frac{\Delta M}{\Delta t} \quad \text{and} \quad \sigma_S = \frac{\Delta M}{\Delta t} \left[ \frac{\sigma_{\Delta M}^2}{(\Delta M)^2} + \frac{\sigma_{\Delta t}^2}{(\Delta t)^2} \right]^{\frac{1}{2}} \tag{10.38}$$

This results in

$$\frac{S}{N} = \frac{S}{\sigma_S} = \left[ \frac{\sigma_M^2}{(\Delta M)^2} + \frac{\sigma_{\Delta t}^2}{(\Delta t)^2} \right]^{-\frac{1}{2}} \tag{10.39}$$

where $M =$ the measured quantity, and $\sigma_S =$ its standard deviation, the rms noise. Equation 10.39 for the case of a fixed-time rate measurement reduces to

$$(S/N)_{\text{fixed } t} = \Delta M_1 (\sigma_{M_1}^2 + \sigma_{M_2}^2)^{\frac{1}{2}} \tag{10.40}$$

where $\sigma_{M1}^2$ and $\sigma_{M2}^2$ are the variances in $M_1$ and $M_2$ respectively. Equation 10.40 is derived from eq. 10.39; in the fixed-time approach $\Delta t$ is constant, and the variance in $\Delta t$ can usually be assumed to be negligible since the reproducibility in time is much superior to the reproducibility of signal measurement.

In the variable-time approach $\Delta M$ is held constant and contemporary switching devices (signal sensors) result in negligible $\sigma_M^2$ values; then the signal-to-noise ratio in the variable-time approach is given by.

$$(S/N)_{\text{variable } t} = \Delta t / (\sigma_{t_1}^2 + \sigma_{t_2}^2)^{\frac{1}{2}} \tag{10.41}$$

For fixed-time spectrophotometric monitoring, Ingle and Crouch (37) concluded that precision is highly dependent on the magnitude of the photocathodic current and source flicker noise as well as on measurement time. Measurements of rates of fast reactions with stopped-flow mixing are limited by photocurrent shot noise. In the measurement of the rates of slow reactions, source flicker noise seems to be the dominant factor in limiting precision. Source stabilization (as discussed in Chapter 9) is one way to solve the problem.

Signal-to-noise expressions for fluorometric kinetic-based determinations have been presented by Wilson and Ingle (38). Using the same lines of reasoning outlined above for fixed- and variable-time procedures, for fluorescence rate measurements (analyte plus background) without manipulation of signal ratios, one can write $S/N$ as follows:

$$S/N = \Delta E_t / \sigma_{(\Delta E)_t}$$
$$= \Delta E_t [mR_f K (E_{f_1} + E_{f_2} + E_{b_1} + E_{b_2})$$
$$+ \xi_1^2 (E_{f_1}^2 + E_{f_2}^2 + E_{b_1}^2 + E_{b_2}^2) + 2\sigma_{dt}^2]^{\frac{1}{2}} \tag{10.42}$$

where the $E$'s identify signal values (in volts): $E_{t_1}$ and $E_{t_2}$ = total signals at times $t_1$ and $t_2$; $E_{b_1}$ and $E_{b_2}$ = background signals at times $t_1$ and $t_2$; $E_{f_1}$ and $E_{f_2}$ = the analyte fluorescence signals at $t_1$ and $t_2$; $\Delta E_t = E_{t_2} - E_{t_1}$; $m$ = the photomultiplier current gain (dimensionless); $R_f$ = the photomultiplier operational amplifier feedback resistor (in ohms); $K$ is the bandwidth constant; $\xi$ = the source flicker factor for photomultiplier output (dimensionless); and $\sigma_{dt}$ = the rms noise including shot and excess noise in dark current and amplifier readout noise (in volts). For a fluorescence rate measurement in which the ratio of outputs is obtained, eq. 10.42 takes the following form:

$$(S/N) = \frac{\Delta E_{tr}}{\sigma_{(\Delta E)_{tr}}} = \left[ \frac{mR_f K(E_{f_1} + E_{f_2} + E_{b_1} + E_{b_2})}{\Delta E_t} \right.$$
$$+ \frac{\xi_2^2(E_{f_1}^2 + E_{f_2}^2 + E_{b_1}^2 + E_{b_2}^2) + 2\sigma_{dt}^2}{\Delta E_t^2}$$
$$\left. + \frac{2R_f' K' E_r + 2\xi_2^2 E_r^2 + 2(\sigma_{dt}')^2}{E_r^2} \right] \tag{10.43}$$

where $E_r$ = reference voltage (phototube).

Experimental data indicates that $S/N$ is about a factor of 2 better when the ratio of signals is utilized. If sample introduction and imprecisions in mixing are not limiting, such a ratio would provide greater advantage in measurements in which source flicker noise, rather than signal shot noise, is limiting.

A comparison of reaction rate methods for systems operating under first-order or pseudo-first-order kinetics has been offered by Wentzell and Crouch (39). Table 10.3 summarizes the precision of the methods compared by these authors under conditions in which the rate coefficient is assumed to remain

**Table 10.3. Relative Standard Deviations (RSD) as a Comparison of Precision of Several Reaction Rate Methods under Conditions of Invariant Rate Coefficient**

| Method | Parameters | RSD(%) |
|---|---|---|
| Derivative | $t = 0.5\,\tau$ | 0.57 |
| Optimized derivative | $t = \tau$ | 1.1 |
| Two-rate | $t_1 = 0.5\,\tau,\ t_2 = 1\,\tau$ | 1.1 |
| Two-rate | $t_1 = 0.5\,\tau,\ t_2 = 1.76\,\tau$ | 0.86 |
| Fixed-time | $t_1 = 0,\ t_2 = 1\,\tau$ | 1.2 |
| Fixed-time | $t_1 = 0,\ t_2 = 2\,\tau$ | 0.83 |
| Optimized fixed-time | $t_1 = 0.8\,\tau,\ t_2 = 1.23\,\tau$ | 5.0 |
| Optimized fixed-time | $t_1 = 0.5\,\tau,\ t_2 = 1.76\,\tau$ | 1.7 |
| Optimized fixed-time (smoothed) | $t_1 = 0.5\,\tau,\ t_2 = 1.76\,\tau$ | 0.41 |

$\tau = 1/k$

*Source*: Reprinted with permission from P. D. Wentzell and S. R. Crouch, *Anal. Chem.*, **58**, 2855. Copyright 1986 by the American Chemical Society.

constant. Optimized methods are those in which measurement is made at $t = 1/k$, and the smoothing of data from the optimized fixed-time method was done with the Savitzky–Golay linear smoothing technique (17). Data in the table confirm the conclusion that could be reached by intuitive thinking— that integral methods are more precise than differential methods. The variable-time approach is much more suitable for zero-order kinetics, and the optimized methods are much less sensitive to between-run changes in rate coefficients. Optimized methods, however, will exhibit poorer precision than traditional methods in the absence of fluctuations in rate coefficients.

### 10.4. KALMAN FILTERING

Linear and nonlinear least-squares regression techniques can be misused if care is not applied in ascertaining the suitability of the model for describing the data. Pitfalls of regression analysis have been pointed out, for instance, by Stowe and Mayer (40) and more recently by Meites (41). Rutan and Brown (42) have explored the use of *extended Kalman filtering*, which is supposed to give results similar to those of regression analysis (42). The approach allows kinetic parameters to be obtained from noisy data.

The Kalman filter, another type of nonlinear regression procedure, is an algorithm for recursive estimation of parameters and states from noisy measurements and has been extensively applied in engineering (43, 44). The nonlinear form of the filter is called the extended Kalman filter and has been used to extract kinetic parameters from a nonlinear model of linear scan voltammograms (45). The extended Kalman filter is based on a Taylor's series expansion about the current estimate of the parameters, with truncation after the linear term. It requires a model given by two equations describing the dynamics of the system and the measurement process. For nonlinearity the model equations are

$$X(j) = f[X(j), X(j-1)] + w(j) \tag{10.44}$$

$$Z(j) = h^T[X(j)] + v(j) \tag{10.45}$$

where $X(j)$ = the vector containing the states to be estimated, $f[X(j), X(j-1)]$ = a nonlinear function describing propagation of these states in time, $h^T[X(j)]$ = a nonlinear function describing the measurement, and $w(j)$ and $v(j)$ = noise contributions; $Z_j$ = measurement made (e.g., absorbance at time $t_j$) (42). In estimating kinetic parameters for reaction $A + B \xrightarrow{k} P$ run under first-order or pseudo-first-order conditions, the state vector is defined by

$$X = \begin{bmatrix} [A]_0 \\ [B]_0 \\ k \end{bmatrix}$$

Table 10.4 summarizes algorithm equations for application of the extended Kalman filter. The equations in this table are used to calculate the state and covariance estimates. Given monitoring of a signal directly proportional to concentration (e.g., absorbance or amperometric current) and all reactants and products contributing in some manner to the signal value, the function $h^T[X(j)]$ can be obtained from

$$h^T[X(j)] = [\rho_A(j) + \rho_B(j) - \rho_P(j)] [A]_0 e^{-kt} + \rho_B(j)[B]_0$$
$$+ [A]_0 [\rho_P(j) - \rho_B(j)] \tag{10.46}$$

The derivatives forming vector $H^T(j)$ are then given by

$$H_1(j) = [\rho_A(j) + \rho_B(j) - \rho_P(j)] e^{-kt} + [\rho_P - \rho_B(j)] \tag{10.47}$$

$$H_2(j) = \rho_B \tag{10.48}$$

$$H_3(j) = -t_j[\rho_A(j) + \rho_B(j) - \rho_P(j)] [A]_0 e^{-kt} \tag{10.49}$$

The elements $H^T$ depend on the current estimates for the parameters (eqs. 10.47, 10.48, and 10.49), and whether or not the parameters are observable

**Table 10.4. Algorithm Equations for Extended Kalman Filter and Definitions of Corresponding Variables**

State estimate extrapolation:    $\mathbf{X}(j|j-1) = \mathbf{X}(j-1|j-1)$

Error covariance extrapolation:    $\mathbf{P}(j|j-1) = \mathbf{P}(j-1|j-1) + \mathbf{Q}(j)$

Kalman gain:    $\mathbf{K}(j) = \mathbf{P}(j|j-1)\cdot\mathbf{H}(j)[\mathbf{H}^T(j)\cdot\mathbf{P}(j|j-1)\cdot\mathbf{H}(j) + \mathbf{R}(j)]^{-1}$

State estimate update:    $\mathbf{X}(j|j) = \mathbf{X}(j|j-1) + \mathbf{K}(j)\{Z(j) - h^T[\mathbf{X}(j|j-1)]\}$

Covariance update:    $\mathbf{P}(j|j) + [\mathbf{I} - \mathbf{K}(j)\cdot\mathbf{H}^T(j)]\cdot\mathbf{P}(j|j-1)$

where

$\mathbf{Q}(j)\ = \mathbf{E}[\mathbf{w}(j)\mathbf{w}^T(i)]\,\delta_{ij}$ $(i,j = 0$ for static systems)

$\mathbf{R}(j)\ = \mathbf{E}[v(j)v(i)]\,\delta_{ij}$

$$\mathbf{H}^T(j) = \left\{\frac{\partial h^T[\mathbf{X}(j|j-1]}{\partial X_1}\ \frac{\partial h^T[\mathbf{X}(j|j-1)]}{\partial X_2}\ \frac{\partial h^T[\mathbf{X}(j|j-1)]}{\partial X_3}\right\}$$

*Definitions of variables*:

$\mathbf{X}(j)$    vector of state variables $(n \times l)$

$\mathbf{H}(j)$    linearized measurement function vector $(n \times l)$

$\mathbf{w}(j)$    system noise vector $(n \times l)$

$v(j)$    measurement noise

$\mathbf{I}$    identity matrix

$\mathbf{Q}(j)$    system covariance matrix $(n \times m)$

$\mathbf{R}(j)$    measurement variance

$\mathbf{P}(j)$    error covariance matrix $(n \times m)$

$m$    number of points filtered

$n$    number of parameters

$j$    index for measurement or estimate at time $t_j$

*Source*: Ref. 42. Reproduced with permission of Elsevier Scientific Publishing Co.

depends on the initial estimates for the parameters initiating the filter. The information matrix is given by

$$\mathbf{I}_n = \mathbf{R}^{-1}\sum_{l=1}^{m}\begin{bmatrix} H_1(l)H_1(l) & H_1(l)H_2(l) & H_1(l)H_3(l) \\ H_1(l)H_2(l) & H_2(l)H_2(l) & H_2(l)H_3(l) \\ H_1(l)H_3(l) & H_2(l)H_3(l) & H_3(l)H_3(l) \end{bmatrix} \qquad (10.50)$$

in which $\mathbf{R}$ is the variance in the measurement process and $m$ is the number of points used to estimate the parameters. The largest square matrix contained within the information matrix that has a nonzero determinant establishes which of the parameters is observable. Practical applications of the extended Kalman filter to kinetic analytical problems are given by Rutan and Brown (42) and Brown et al. (45). Rutan and Brown (42), for instance, report successful use of the extended Kalman filter in (a) estimating the rate coefficient for the dissociation of the 1:1 praseodymium–DCTA complex, and (b) a demonstration of how to prevent the use of an erroneous model (chemical example used: aquation of *trans*-$CrCl_2^+$). The use of the Kalman filter to estimate the final absorbance from data collected before the reaction has gone to completion, and its usefulness in analyzing multicomponent reacting systems, was discussed by Crouch during the Second International Symposium on Kinetics in Analytical Chemistry (46).

The Kalman filter has been discussed by Cooper (47) within the context of using optimal estimation theory in data analysis and signal processing. Brown (48) reviewed Kalman filter applications in analytical chemistry and Rutan (49) discussed in detail Kalman filtering in multicomponent curve resolution and concentration estimation, correction for variable background responses, calibration with drift compensation, and estimation of kinetic parameters for first-order reactions and for heterogeneous charge-transfer processes. A recent contribution to the application of extended Kalman filtering is the demonstration of rate coefficient determination while the reaction temperature is changing (50).

## REFERENCES

1. T. E. Hewitt and H. L. Pardue, *Clin. Chem. (Winston–Salem, N.C.)*, **19,** 1128 (1973).
2. P. W. Carr, *Anal. Chem.*, **50,** 1602 (1978).
3. J. B. Landis, M. Rebec, and H. L. Pardue, *Anal. Chem.*, **49,** 785 (1977).
4. F. J. Holler, R. K. Calhoun, and S. F. McClanahan, *Anal. Chem.*, **54,** 755 (1982).
5. J. G. Atwood and J. L. DiCesare, *Clin. Chem. (Winston–Salem, N. C.)*, **19,** 1263 (1973).
6. P. D. Wentzell and S. R. Crouch, *Anal. Chem.*, **58,** 2851 (1986).
7. J. B. Pausch and D. W. Margerum, *Anal. Chem.*, **41,** 226 (1969).
8. A. C. Javier, S. R. Crouch, and H. V. Malmstadt, *Anal. Chem.*, **41,** 239 (1969).
9. G. E. Mieling and H. L. Pardue, *Anal. Chem.*, **50,** 1611 (1978).
10. D. W. Marquardt, *J. Soc. Ind. Appl. Math.*, **11** (2), 431 (1963).
11. P. R. Bevington, *Data Reduction and Error Analysis for the Physical Sciences*, McGraw-Hill, New York, 1969.
12. H. L. Pardue, *Clin. Chem. (Winston–Salem, N.C.)*, **23,** 2189 (1977).
13. G. E. Mieling, H. L. Pardue, J. E. Thompson, and R. A. Smith, *Clin, Chem. (Winston–Salem, N. C.)*, **25,** 1581 (1979).
14. R. S. Harner and H. L. Pardue, *Anal. Chim. Acta*, **127,** 23 (1981).

15. J. D. Lin and Harry, L. Pardue, *Clin. Chem. (Winston–Salem, N. C.)*, **28**, 2081 (1982).

16. S. D. Hamilton and H. L. Pardue, *Clin. Chem. (Winston–Salem, N. C.)*, **28**, 2359 (1982).

17. A. Savitzky and J. E. Golay, *Anal. Chem.*, **36**, 1627 (1964).

18. R. C. Harris and E. Hultman, *Clin. Chem. (Winston–Salem, N. C.)*, **29**, 2079 (1983).

19. C. Smith, *London Edinburgh Philos. Mag. J. Sci.*, **1**, 496 (1926).

20. M. P. Goren and J. E. Davis, *Clin. Chem. (Winston–Salem, N. C.)*, **32**, 2021 (1986).

21. R. L. Czervionke, G. F. Johnson, N. F. Nuwayhid, and R. D. Feld, *Clin. Chem. (Winston–Salem, N. C.)*, **32**, 1648 (1986).

22. H. U. Bergmeyer, *Biochem. Z.*, **323**, 408 (1953).

23. J. E. Davis and J. Pevnick, *Anal. Chem.*, **51**, 529 (1979).

24. H. A. Mottola, *J. Chem. Educ.*, **58**, 399 (1981).

25. C. H. Denbow and V. Goedicke, *Foundations of Mathematics*, Harper & Row, New York, 1959, pp. 482–486.

26. L. J. Papa, H. B. Mark, Jr., and G. A. Rechnitz, "Evaluation and Comparison of Differential Rate Methods Employed for the Simultaneous Determination of Closely Related Compounds," in *Kinetics in Analytical Chemistry*, H. B. Mark and G. A. Rechnitz, Eds., Interscience, New York, 1968, Chapter 7.

27. G. M. Ridder and D. W. Margerum, "Simultaneous Kinetic Analysis," in *Analytical Chemistry: Essays in Memory of Anders Ringbom*, E. Wanninen, Ed., Pergamon, New York, 1977, pp. 515–528.

28. B. G. Willis, W. H. Woodruff, J. M. Frysinger, D. W. Margerum, and H. L. Pardue, *Anal. Chem.*, **42**, 1350 (1970).

29. N. R. Draper and H. Smith, *Applied Regression Analysis*, Wiley, New York, 1966.

30. W. E. Wentworth, *J. Chem. Educ.*, **42**, 96 (1965); **42**, 162 (1965).

31. B. G. Willis, J. A. Bittikofer, H. L. Pardue, and D. W. Margerum, *Anal. Chem.*, **42**, 1340 (1970).

32. I. A. G. Roos, L. P. G. Wakelin, J. Hakkennes, and J. Coles, *Anal. Biochem.*, **146**, 287 (1985).

33. J. C. Thompsen and H. A. Mottola, *Anal. Instrum.*, **13**, 89 (1984).

34. J. W. Frazer and H. R. Brand, *Anal. Chem.*, **52**, 1730 (1980).

35. C. A. Burtis, W. D. Bostick, J. B. Overton, and J. E. Mrochek, *Anal. Chem.*, **53**, 1154 (1981).

36. J. D. Ingle, Jr., and S. R. Crouch, *Anal. Chem.*, **43**, 697 (1971).

37. J. D. Ingle, Jr., and S. R. Crouch, *Anal. Chem.*, **45**, 333 (1973).

38. R. L. Wilson and J. D. Ingle, Jr., *Anal. Chem.*, **49**, 1060 (1977).

39. P. D. Wentzell and S. R. Crouch, *Anal. Chem.*, **58**, 2855 (1986).

40. R. A. Stowe and R. P. Mayer, *Ind. Eng. Chem.*, **61,** 11 (1969).

41. L. Meites, *CRC Crit. Rev. Anal. Chem.*, **8,** 1 (1979).

42. S. C. Rutan and S. D. Brown, *Anal. Chim. Acta*, **167,** 23 (1985).

43. A. Gelb, Ed., *Applied Optimal Estimation*, MIT Press, Cambridge, Mass., 1974.

44. A. H. Jazwinski, *Stochastic Process and Filtering Theory*, Academic Press, New York, 1970.

45. T. F. Brown, D. M. Caster, and S. D. Brown, *Anal. Chem.*, **56,** 1214 (1984).

46. S. R. Crouch and P. D. Wentzell, *Lecture L–1, Abstr. 2nd Int. Symp. Kinetics in Anal. Chem.*, Preveza, Greece, September 1986.

47. W. S. Cooper, *Rev. Sci. Instrum.*, **57,** 2862 (1986).

48. S. D. Brown, *Anal. Chim. Acta*, **181,** 1 (1986).

49. S. C. Rutan, *J. Chemometrics*, **1,** 7 (1987).

50. S. C. Rutan and C. A. Corcoran, Abstracts of the 1987 Pittsburgh Conference and Exposition, Abstract No. 550, Atlantic City, NJ, March 9–13, 1987.

# CHAPTER

## 11

# KINETIC COMPONENTS IN SEVERAL ANALYTICAL
# TECHNIQUES OR STEPS IN ANALYSIS

The bulk of the material in this monograph, reflecting the overwhelming literature coverage, is devoted to kinetic-based methods of determination. However, the roles played by kinetics in several analytical techniques or steps in analysis should not be overlooked. This chapter is intended to bring balance by describing these roles played by fundamental kinetic aspects in contemporary analytical chemistry.

Many contemporary techniques in analytical methodology offer opportunities to note the relevance of kinetics because of the dynamic conditions under which such techniques or steps operate. Examples from (a) chromatographic separations, (b) nonflame (furnace) atomic absorption determinations, and (c) continuous-flow methods were pointed out some years ago (1) with reference to teaching quantitive analysis at the college level. As pointed out at that time, textbooks and courses usually emphasize procedures, techniques, and instruments for determinative purposes, and many fundamental concepts tend to get diluted or ignored. Unfortunately, I must report that more recent textbooks on so-called quantitative chemical analysis have shown no improvement in correcting such neglect. Exposing students in analytical courses (undergraduate as well as first-year graduate level) to these concepts should provide them with a better grasp of the fundamentals behind analytical techniques, put kinetics in analytical chemistry into better perspective, and open the mind of the student to a level of thinking needed to promote creativity.

Table 11.1 lists typical steps that define a particular analysis (4). Many aspects of those steps, if scrutinized closely, will be seen to involve a kinetic component. Kinetics impacts on many aspects of sample collection and transport to analysis (sampling). Rate of oxidation (e.g., by dissolved oxygen) can alter the oxidation state of many transition metal ions in water samples to be analyzed for pollutants. The same can be said of the rate of hydrolysis of some chemical species. In general, chromatography is a totally kinetically controlled process, and as indicated in the table it constitutes the bulk of contemporary separation steps. As documented in previous chapters, many

236

**Table 11.1.   Typical Steps in a Given Analysis**

| Step | Remarks |
| --- | --- |
| Sampling | Mainly physical operations of sample collection guided by statistical considerations. Unfortunately only token mention is given in the training of analytical chemists at colleges and universities. Its inclusion as a separate topic in a recent fundamental review (2) may indicate increased awareness of its relevance in the overall picture of chemical analysis. |
| Sample preparation | Includes physical and chemical operations to prepare the sample for adequate manipulations in subsequent steps (crushing to reduce to powder, dissolution in an appropriate solvent, dilution, chemical derivatization, etc.) |
| Separations and/or Preconcentrations | The bulk of the work in these areas in contemporary analysis involves chromatography and solvent extraction. Only occasionally are traditional operations such as precipitation or distillation part of this step. |
| Measurement | Commonly identified as the determination step, although some redefinitions have been proposed (3). The bulk of training given at colleges and universities focuses on this step. The measured properties are generally physical in nature (e.g., absorption or emission of radiant energy, conductivity, electrochemical electron transfer), but their changes in numerical value are often the result of chemical reactions introduced to gain selectivity.<br><br>These measurements may be performed under equilibrium conditions (thermodynamic methods of determination) or nonequilibrium conditions (kinetic methods). |
| Data manipulation | This step is dominated by electronics and use of computers. From a quantitative viewpoint it is oriented to provide as final information the most probable value for a given quantity and its associated uncertainty. As such, statistical mathematics is at the heart of this step. Since most contemporary analytical measurement approaches are based on standardization and |

**Table 11.1**   (*continued*)

| Step | Remarks |
|------|---------|
| | linear calibration curves, best-fit regression treatments are generally facilitated by the use of computers. |
| | This automation step makes heavy use of analog-to-digital and digital-to-analog conversion in the electrical domain and final transformation to numerical (or graphical) form in the nonelectrical domain. |
| Interpretation of results | Although frequently analytical chemists hand on the results of their work to other scientists for interpretation, involvement in fundamental and/or interdisciplinary research by the analytical scientist makes this step part of the analysis as a whole. |

*Source*: From Ref. 4.

measurements are performed under dynamic conditions and physical as well as chemical kinetics dominate their performance. Illustrative examples discussed in some detail are presented in the following sections.

## 11.1.   DIFFUSION IN ANALYTICAL CHEMISTRY

A myriad of analytical steps take place with transport of mass from one location to another. Mass transport on a molecular scale is called *diffusion*, and bulk motion (movement of the system as a whole) is known as *convection*. In many cases, such as in turbulent eddies, transport on an intermediate scale needs to be considered. Diffusion is perhaps the most important mechanism of mass transport in analytical methodology. Many chemical reactions in solution have elementary steps that are diffusion-controlled (5). Diffusion is fundamental in separation procedures (chromatography in particular) (6). Diffusion-controlled electrochemical processes are of capital importance in electroanalytical chemistry (7), and diffusion is at the heart of voltammetric techniques. Axial and radial diffusion greatly contribute to dispersion in unsegmented continuous-flow sample–reagent(s) processing (8). These facts justify presenting here the basic tools describing diffusion since it represents an important physical kinetic process in a variety of topics of analytical relevance.

The most common approach to describe diffusion, which occurs any time there is a concentration difference (gradient) between regions of a system, is Fick's first and second laws (9). They resulted from the efforts of Adolf Eugen Fick (1829–1901) to apply the principles of physics to physiology (Fick was professor of physiology at the University of Würzburg, Germany).

Diffusion drives the system in the direction that will equalize the concentration value throughout the system by moving some of the molecules (ions) from the higher to the lower concentration region. *Fick's first law of diffusion* gives a means of determining the transport of a given amount of material, $W$, across a given area $A$ as the result of a concentration gradient $\partial C/\partial x$ in the direction $x$ normal to the plane containing area $A$, during time $t$:

$$W = -DA(\partial C/\partial x)t \qquad (11.1)$$

where $D$ is the proportionality constant known as the *diffusion coefficient*. The negative sign is introduced so that $D$ will have a positive value. The diffusion coefficient represents the amount of solute that diffuses across a unit area in 1 s under the influence of unit concentration gradient.

*Fick's second law of diffusion* in its one-dimensional form,

$$\partial C/\partial t = -D(\partial^2 C/\partial x^2) \qquad (11.2)$$

can be written in the generalized manner

$$dC/dt = D\nabla^2 C \qquad (11.3)$$

where $\nabla^2$ = the differential operator or Laplacian operator. $\nabla^2 C$ is a measure of how much the value of $C$ at $(x, y, z)$ differs from the average of the values of $C$ at surrounding points, and it can be written as

$$\nabla^2 C = (\partial^2 C/\partial x^2) + (\partial^2 C/\partial y^2) + (\partial^2 C/\partial z^2) \qquad (11.4)$$

Equation 11.2 indicates that the change in concentration with time is related to the diffusion coefficient and the slope of the concentration gradient.

Fick's first law is empirical in essence and describes an idealized situation. The diffusion coefficient may vary with concentration, and the chemical potential gradient, rather than the concentration gradient, is the actual driving force. However, the mathematical equation is valid in dilute solution. Many diffusional problems in analytical chemistry can be treated as unidirectional (in a preferred direction), although in actuality diffusion can take place in any direction in which a concentration gradient is present. Multidirectional diffusion is of interest in analytical steps in which mixing is the goal (e.g., enhancement of radial diffusion and some degree of axial diffusion in the sample "plug" and its boundaries in unsegmented continuous-flow systems in which chemical reactions take place between components of the sample and the carrier stream; see Chapter 9). Unidirectional diffusion, on the other hand, is of interest in separation steps.

Fick's first law is, however, inadequate for solving most problems in diffusion since concentration gradients are actually unknown.

The second-order partial differential equation of Fick's second law (eq. 11.2) can be solved relatively easily for a wide variety of boundary conditions and is the basic mathematical tool considered to govern diffusion problems in analytically relevant cases.

## 11.2. KINETICS IN CHROMATOGRAPHY

Chromatography provides an interesting topical example. Because of the nature of the chromatographic process, not all the molecules entering a column at time $t = 0$ will leave the column at the same time. The residence-time distribution curve (chromatogram) is obtained by plotting versus time the fraction of molecules having a residence time between $t$ and $t + dt$. The average residence time is called the retention time. The classical *plate theory* (10) derived from thermodynamic equilibrium, never attained in contemporary chromatography, has provided useful formulations for practical calculation of retention data. In doing so, however, the plate theory has many times obscured the fact that it fails to give a good account of band spreading; moreover, to unaware students, it gives the wrong impression that chromatography is in essence an equilibrium-based process. The introduction of the rate theory of chromatography by Van Deemter et al. (11) put the basis of the chromatographic process in better perspective, and the generalized nonequilibrium theory (12) has provided a clearer understanding of the factors (kinetic in nature) contributing to band spreading.

It is of interest to quote Van Deemter et al. (11):

The rate theory in principle provides all information on the influence of kinetic phenomena such as rate of mass transfer between phases, rate of adsorption or chemical reaction, longitudinal diffusion and flow behavior on the history of a band in the column.

Chromatographic zone spreading is generally described by two approaches: (a) the material- (or mass-) balance approach, and (b) the random-walk model. The general mass-balance equation, describing concentration changes, takes the form (13)

$$\partial C/\partial t = D(\partial^2 C/\partial x^2) - v(\partial C/\partial x) + \sum_j (k_{ij} C_j - k_{ji} C_i) \qquad (11.5)$$

in which $k_{ij}$ and $k_{ji}$ = the first-order rate constants for the interconversion

$$S_i \underset{k_{ji}}{\overset{k_{ij}}{\rightleftharpoons}} S_j$$

S = a solute molecule on a particular type of site, and $i$ and $j$ = states that can be reached (e.g., by a solute species) through first-order processes (e.g., mass transfer between phases). $D$ = the diffusion coefficient, $x$ = the distance traveled in the direction of flow, $C$ = the instantaneous concentration value, and $v$ = the instantaneous flow velocity. On the right-hand side of eq. 11.5 the first term accounts for unidirectional diffusion effects, the second term for convection, and the third for phase-transfer kinetics. As such, the description of the chromatographic process via its more important aspect, band broadening, and in a general manner is totally kinetic in nature. Solution of eq. 11.5 yields the solute distribution function.

In a simplified manner it is helpful to associate the ideal chromatogram with a Gaussian curve (although a Poisson distribution is more realistic), the spread of which is kinetically controlled by diffusion and Fick's second law. Direct integration of eq. 11.2 for diffusion from a planar source at constant flow rate along the bed leads to

$$C = \frac{1}{2(\pi D t)^{\frac{1}{2}}} \exp\left(\frac{-x^2}{4Dt}\right) \tag{11.6}$$

Expression of this integrated form in terms of the standard deviation $\sigma$ of the Gaussian shape assumed by the chromatographic zone as it migrates along the column is

$$C = \frac{1}{\sigma (2\pi)^{\frac{1}{2}}} \exp\left(\frac{-x^2}{2\sigma^2}\right) \tag{11.7}$$

Comparing 11.6 and 11.7 gives

$$\sigma = (2Dt)^{\frac{1}{2}} \tag{11.8}$$

Equation 11.8 shows the kinetic nature of band spreading, since the standard deviation is directly proportional to a rate coefficient (the diffusion coefficient) and to the time elapsed since penetration of a "population" of molecules into the corresponding phase.

Equation 11.8 also permits a better appreciation of kinetics in the random-walk model. The basic assumption for the model is that a solute moves through the chromatographic column with its molecules undergoing a large number of sorption–desorption steps in an entirely random manner. If, for simplicity, only one dimension is considered and a large number of molecules take a random walk of $N$ steps of length $l$, the spreading with time can be expressed as

$$\sigma = l(N)^{\frac{1}{2}} \tag{11.9}$$

Many random walks of different $N$ and $l$ are simultaneously possible, and the combined effect in band spreading can be given as the sum of the corresponding variances:

$$\sigma^2 = \sigma_1^2 + \sigma_2^2 + \sigma_3^2 + \cdots + \sigma_i^2 \tag{11.10}$$

Chromatographically, $l$ corresponds to the "plate height" $H$ and the product $lN$ to the distance migrated in the direction of $x$, $L_x$. Then

$$\sigma^2 = l^2 N = HL_x \tag{11.11}$$

From this the kinetic connotation of plate height can be realized immediately, since

$$H = \sigma^2 / L_x$$

## 11.3. KINETICS IN ELECTROCHEMICAL PROCESSES

The general series of steps that may be involved in the conversion of a dissolved oxidized species to its reduced form, or vice versa, is illustrated in Figure 11.1. The electrode reaction rate can be effected then by

1. Mass transfer.
2. Electron transfer at the electrode surface.
3. Chemical reactions preceding or following the electron transfer at the electrode surface.
4. Surface interactions such as adsorption, desorption, or electro-deposition.

Each of these steps has its own characteristic kinetics. Mass transfer, for instance, can be effected by (a) migration (movement of a charged body under the influence of a gradient of electrical potential), (b) diffusion, or (c) convection. The mass transfer to the electrode surface is governed by the *Nernst–Planck* equation, which for a unidimensional mass transfer along the

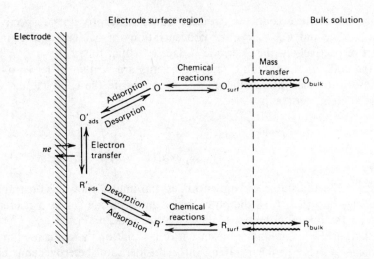

**Figure 11.1.** Possible chemical and electrochemical steps in a general electrode reaction of the type $O + ne \rightleftharpoons R$.[From Ref. 7.]

$x$ axis can be written as

$$J_i(x) = -D_i\frac{\partial C_i(x)}{\partial x} - \frac{z_i\mathscr{F}}{RT}D_i C_i\frac{\partial \phi(x)}{\partial x} + C_i v(x) \qquad (11.12)$$

where $J_i(x) =$ the flux of species $i$ (in mol/s·cm²) at distance $x$ from the surface, $\phi =$ the electrical potential, $z =$ the charge of species $i$, $\mathscr{F} =$ the Faraday constant, and $v(x) =$ the velocity (in centimeters per second) with which a volume element in solution moves along $x$. The three terms on the right-hand side of eq. 11.12 represent the contributions of diffusion, migration, and convection, respectively.

If the solution is well stirred (i.e., transport of electroactive species to the electrode surface is fast) and neither chemical reactions prior to (or after) the electron exchange nor surface interactions take place, one can focus on the kinetics of the electron-transfer process only (14). For the generalized elementary electron transfer process of Figure 11.1 and with the assumption of first-order oxidation and reduction processes, the difference between the cathodic and anodic rates is the net flux of O at the electrode surface and is proportional to the net current $i$ at time $t$:

$$i = \mathscr{F}A[k_c C_{O(x=0,t)} - k_a C_{R(x=0,t)}] \qquad (11.13)$$

where $\mathscr{F} =$ the Faraday constant, $A =$ the electrode area, $k_c$ and $k_a$ the rate coefficients characterizing the cathodic and anodic processes respectively, and $C_{O(x=0,t)}$ and $C_{R(x=0,t)}$ = the concentrations of oxidized and reduced species respectively at the electrode surface ($x=0$) at time $t$.

If the cathodic and anodic rate coefficients are written in terms of the standard (intrinsic) rate coefficient $k^\circ$, we can write the complete current-potential characteristic (7):

$$i = n\mathscr{F}A\,k^\circ \{C_{O(x=0,t)}\, \exp[-\alpha nf(E-E^{\circ\prime})]$$
$$- C_{R(x=0,t)}\, \exp[(1-\alpha)nf(E-E^{\circ\prime})]\} \qquad (11.14)$$

where $\alpha$ = the adjustable and dimensionless parameter known as the transfer coefficient, $f=\mathscr{F}/RT$, $E$ = the potential of the electrode versus a reference, and $E^{\circ\prime}$ = the formal potential of the same electrode.

Equation 11.14 is fundamental since it or a variation thereof is used in the treatment of every problem taking into consideration heterogeneous electrode kinetics (7). The standard rate coefficient $k^\circ$ is a measure of the sluggishness of the kinetics of a redox couple [i.e., the electrochemical reversibility of the system (15)]. A large value of $k^\circ$ indicates attainment of equilibrium in a short time (perfectly reversible system), so that the concentration values of O and R are dictated by the Nernst equation. In a system with a small value of $k^\circ$, however, the electron-transfer step at the electrode surface will be slow (irreversible system). In practice, of course, the observation of reversibility depends not only on the value of $k^\circ$ but also on the time scale of the device monitoring the concentration of O and/or R. The largest measured standard rate coefficients are in the range of 1 to 10 cm/s and correspond to simple electron-transfer processes.

The overall picture of kinetics in electrochemical processes of analytical interest is completed by realizing that pre- or postelectron-transfer chemical reactions as well as surface interactions may become relevant to the overall rate of the process.

## 11.4.  KINETICS IN ABSORPTION/EMISSION SPECTROSCOPY

The intrinsic kinetic nature of light-emitting processes (e.g., fluorescence, phosphorescence, and chemi- and bioluminescence) were considered in Chapter 8 and are cited here only for reference.

The operating characteristics of certain atomic absorption processes result in transient signals that can be described kinetically. Such a description is of obvious analytical interest, as is the kinetic dependence of steady-state signals in flame emission spectroscopy.

### 11.4.1.   Series Processes in Nonflame Atomic Absorption Spectroscopy

The use of a furnace in atomic absorption spectroscopy was first introduced by L'vov (16, 17). Transient signals produced by single-shot sampling devices have become common with the introduction of heated graphite-tube atomizers (18, 19). In four papers Fuller (20) has presented and discussed a consecutive (series) kinetic model of interest in furnace atomic absorption spectroscopy.

The model can be summarized as follows:

$$M(\text{unvaporized}) \xrightarrow{k_1} M(g) \text{ (in furnace)} \qquad (11.15)$$

$$M(g) \text{ (in furnace)} \xrightarrow{k_2} M(g) \text{ (out of furnace)} \qquad (11.16)$$

Here M represents atoms of a given metallic element. Fuller developed and tested his kinetic model using copper as M since the production of copper atoms can be measured conveniently at relatively low temperatures (1700 to 2200 K), copper does not form a stable carbide, and it has relatively low melting and boiling points, which should simplify the kinetics of atom formation. If each step in eqs. 11.15 and 11.16 is first-order (most pseudo-first-order processes can be similarly treated), the kinetic description leads to easily solved linear differential equations (21). These permit the description of the instantaneous absorbance values in the analyte peak by the expression

$$A_{Cu} = k_1/(k_2 - k_1)\rho[\text{Cu}]_0(e^{-k_1 t} - e^{-k_2 t}) \qquad (11.17)$$

where $A$ = absorbance, $[\text{Cu}]_0$ = the initial concentration of copper present, and $\rho$ = a constant relating the measured absorbance with the amount of copper atoms in the furnace. The first paper in Ref. 20 presents the fundamentals of the kinetic model showing how to obtain (from experimental data) values for the rate coefficients $k_1$ and $k_2$ and the parameter $\rho$. The good agreement between experiment and model is illustrated in Figure 11.2. The studies led to the conclusion that atomization occurs through carbon reduction and to the postulation that the most probable mechanism is

$$\text{Cu}_2\text{O} + \text{C} \xrightarrow{\text{slow}} 2\text{Cu}(s/l) + \text{CO}$$

$$\text{Cu}(s/l) \xrightarrow{\text{fast}} \text{Cu}(g)$$

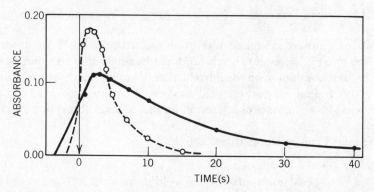

**Figure 11.2.** Characteristic absorbance–time profiles for the atomization of copper. ——— Experimental curve at 1860 K, ● theoretically predicted data points, – – – – experimental curve at 2220 K, ○ theoretically predicted data points. Both curves are for $2.0 \times 10^{-8}$ g Cu. The portion of the curve to the left of the time zero mark corresponds to the initial few seconds of copper production when the temperature of the graphite furnace is still increasing to its preset value. [Adapted from C. W. Fuller, *Analyst (London)*, **99**, 739 (1974), with permission.]

An interesting conclusion of analytical relevance was derived from these studies: a tantalum furnace produces greater sensitivity than one of graphite, at least for the determination of copper. The series process model was used by Fuller (20) to (a) derive an equation to extract parameters encountered in atomic absorption determinations providing a quantitative description of preatomization heating losses, (b) ascertain advantages of operating under stopped-gas-flow conditions, (c) consider the relative merits of measuring signal peak height and signal integration, (d) select atomization parameters for matrix control, and (e) assess interference effects that occur during the atomization period.

### 11.4.2. Other Kinetic Considerations of Interest in Flameless Atomization

Holcombe et al. (22) analyzed interference from gas-phase reactions in flameless atomic absorption. To do so, they used a kinetic model based on the forward rate of the two most likely gas-phase reactions capable of depleting a free-atom population:

$$M + AB \rightleftharpoons MA + B \qquad (11.18)$$

$$M + C + X \rightleftharpoons MC + X \qquad (11.19)$$

where M = the metal species to be determined; A, B and C = interfering species; and X = a third body (e.g., argon gas) that facilitates reaction 11.19.

They concluded that at the partial pressures encountered in systems of this type, reaction kinetics may play a significant role in predicting the extent to which M and the interferent species will interact.

The determination of traces of silicon by furnace atomic absorption spectrometry has poor sensitivity, which is considered to be the result of silicon carbide formation. Muller-Vogt and Wendl (23) carried out experiments in both untreated and niobium-coated graphite tubes and used gas-phase reaction kinetics to interpret the observations. These studies, which included scanning electron microanalysis of the tube surface, provided evidence for reduction of $Na_2SiO_3$ during the ashing cycle. Neither carbide formation nor losses of silicon were detected in either tube up to temperatures of 1650°C. The rate of reduction increases and the activation energy decreases in niobium-coated tubes. To explain this they suggested that niobium carbide, in large excess over $Na_2SiO_3$, takes part in the reduction process, for example, by incorporating oxygen into its lattice.

Despite the number of studies devoted to the kinetics of sample evaporation in nonflame atomic absorption, discrepancies prevent the adoption of a general mechanism for the process. Substantial differences are found in the reported values of the activation energy for the evaporation of ions of the same element as a function of temperature. These differences appear as a break or inflection on the Arrhenius plot. L'vov and Bayunov (24) have developed a diffusion-microkinetic model that explains these breaks as well as the differences between activation energies for the low and high temperature ranges. They have also used their model to explain changes in sensitivity and in the extent of matrix effects due to changes in structure and in the state of the graphite surface as a result of aging. Two different evaporation regions are postulated by L'vov and Bayunov: (a) an effusion-kinetic region in which the evaporation rate is characterized solely by the value of the rate coefficient for evaporation, and (b) a diffusion-kinetic region in which the evaporation rate is dictated by the value of the diffusion coefficient of the material in its condensed phase. In a subsequent paper the same model was used to study volatilization of samples in atomizers manufactured from nonporous materials such as pyrolytic carbon or metals (25).

Musil and Rubeska (26) have included interaction of free atoms with the walls of the observation zone as a third process to be considered in the mathematical modeling of free-atom formation and dissipation in electro-thermal atomization.

A theoretical study of the reasons for the appearance of transient absorption signals during the evaporation of $Al_2O_3$ and other oxides difficult to vaporize in graphite furnace atomic absorption has been advanced by L'vov and Savin (27). They postulate autocatalytic reduction of the oxides by gaseous carbide molecules. The carbides would be formed in the presence of carbon when the metal vapor density is sufficiently high. The kinetics of carbothermal reduction of difficult volatile oxides in graphite furnaces is considered in detail in a subsequent paper (28).

### 11.4.3. Kinetics in Analytical Flame Spectroscopy

The kinetics of excitation and de-excitation processes in flames (emission spectroscopy) has received attention in specialized monographs (29).

Assume that a given species M is excited to a state M* by one (or both) of the following processes:

$$M + Z \underset{k_{-1}}{\overset{k_1}{\rightleftharpoons}} M^* + Z \qquad \text{(inelastic collision)}$$

$$M + X + Y \underset{k_{-2}}{\overset{k_2}{\rightleftharpoons}} M^* + XY \qquad \text{(electrothermic recombination)}$$

where X, Y, and Z are atomic or molecular species of the flame gas. Then the ratio of excited state population to ground state population in a stationary state is given by (29)

$$\frac{[M^*]}{[M]} = \frac{k_1[Z] + k_2[X][Y]}{k_{-1}[Z] + k_{-2}[XY]} \qquad (11.20)$$

since $d[M^*]/dt = 0$. Hence even the efficiency of signal measurement under steady state in simple flame emission spectroscopy is kinetically controlled in the de-excitation of M* by spontaneous photon emission.

Little attention has been given to interferences, and most theoretical treatments describing atomization in analytical flame spectroscopy neglect the diffusion and/or kinetic effects on the atom distribution in the flame. A more general theory considering both kinetic effects (diffusion and chemical kinetics) was advanced by Li (30) for the description of chemical interferences on the atomic distribution from a vaporized aerosol particle. The three most relevant types of chemical interference considered were (a) dissociation, (b) combination, and (c) ionization. In a subsequent publication (31) the theory was developed further to study chemical interference in atomization at various dissociation and vaporization rates.

## 11.5. KINETICS IN CONTINUOUS-FLOW SAMPLE–REAGENT(S) PROCESSING

The practice of wet chemistry (which, on account of permanent interest in health and environmentally related problems, continues to be a sizable part of contemporary analytical chemistry) has not been the same since the introduction of continuous-flow analyzers (32). Kinetics is at the heart of such analyzers, and an appreciation of this should lead to improved methodology. Operational characteristics of these forms of sample and reagent(s) handling were discussed in Chapter 9 with reference to mixing by these approaches, and some of their kinetic components were set forth there. Reviews of concepts pertinent to this section are available (33, 34).

Historically, air-segmented continuous-flow sample–reagent(s) processing preceded the unsegmented form. However, both approaches have kinetic connotations of their own and as such are treated separately. A point of interest is that, if chemical reactions are involved in these systems, due to design and operating principles they do not need to reach equilibrium. Because of this and the inherent kinetic nature of flow, these approaches are per se kinetically controlled, yielding signals whose shapes are correspondingly dictated.

### 11.5.1.  Kinetics in Air-Segmented Sample–Reagent(s) Processing Systems

Recognition of the kinetic nature of signal characteristics in air-segmented flow systems has to be credited to Thiers et al. (35). Details of the factors and practical steps that may be taken to improve peak quality and analytical performance can be found in specialized monographs (36). Thiers et al. (35) describe the rise portion of the curve by

$$dS/dt = k(S_{ss} - S_t) \qquad (11.21)$$

where $k$ = a rate proportionality constant, $S_{ss}$ = the signal at the steady-state portion of the profile (see Figure 11.3), and $S_t$ = the signal at time $t$ in the signal–time profile. The fall portion is described by

$$dS/dt = -kS_t \qquad (11.22)$$

since $S_{ss} = 0$ and drops out from eq. 11.21.

Ideally, the peak profile should be similar to the signal illustrated in Figure 11.3a. Two kinds of deformations (both of kinetic nature), however, together result in signal distortion. These deformations can be classified as (a) segmented-stream deformations, and (b) unsegmented-stream deformations. Segmented-type deformations (taking place in the portions affected by air segments) should theoretically lead to a Poisson distribution of molecules in the plug of solution under examination by the detection device. Because a large number of segments are involved in the distribution, the Poisson model

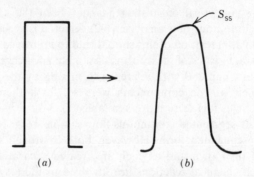

**Figure 11.3.** Signal deformation in segmented continuous-flow systems. (*a*) Ideal square wave signal. (*b*) Signal profile showing the combined deformation as a result of normal probability integral deformation and exponential (unsegmented) deformation.

is indistinguishable from a symmetrical, Gaussian model. The connection between the kinetic nature of this deformation and band broadening in chromatography is obvious. It is no surprise then that the relevant kinetic parameter to evaluate this deformation is the standard deviation of the mean dwell time, $\sigma_D$, and the net effect is termed a *normal probability integral deformation*. The second type of deformation, which is observed mainly in the flow cell and debubbler (used to vent the air bubbles just before detection takes place), under normal conditions takes the form of an exponential curve effect. The relevant kinetic parameter for this kind of deformation is the exponential factor *b*, which can be simply extracted by application of the following relationship:

$$\log S_t = \log \Delta S - (t/b) \tag{11.23}$$

where $\Delta S$ corresponds to the difference in signal (assumed to be directly proportional to the concentration of the monitored species) between initial and final steady-state signals. If a rise or fall curve is plotted semilogarithmically versus time, after a nonlinear portion the rest of the plot is linear and described by eq. 11.23. The early nonlinear part represents the time required for the segmented stream to reach a new steady state at the debubbler. The exponential constant *b* defines the magnitude of the exponential deformation and is graphically defined as the time required to pass from any signal value $S_t$ to $0.37\,S_t$.

Measurement of $\sigma_D$ from a rise or fall curve requires prior removal of the exponential deformation. The process, called curve regeneration (37), is based on the fact that any signal value at the debubbler, $S_t^{db}$ and the corresponding signal after exponential deformation, $S_t$, is given by (38).

$$S_t^{db} = b\,(dS/dt) + S_t \tag{11.24}$$

Signal measurement on a rise or fall curve every 2 s allows one to obtain an estimate of $dS/dt$ by using

$$0.5(S_n + S_{n+1}) \tag{11.25}$$

where $S_n$ and $S_{n+1}$ = consecutive 2-s signal readings. $S_t$ is the midpoint of the two readings, $0.5(S_n + S_{n+1})$. The signal at the debubbler, $S_t^{db}$, is transformed by

$$(S_t^{db})' = (S_t^{db} - S_{min})/(S_{max} - S_{min}) \tag{11.26}$$

in order to prepare probability plots (39) and extract the value of $\sigma_D$. Figure 11.4 shows a plot of the transformed signal at the debubbler on a probability

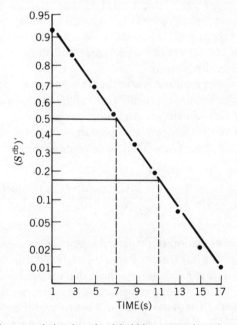

**Figure 11.4.** Plot of corrected signal at the debubbler versus time. A cumulative probability scale is used for the signal and a linear scale for time. The time interval corresponding to a change in $S^{db}$ from 0.5 to 0.16 gives $\sigma_D = 4$ s. [From Ref. 36. Reproduced with permission of Marcel Dekker Inc.]

scale. The transformation by eq. 11.26 is necessary because plots on probability paper have limits of 0 and 1.

The values of $\sigma_D$ and $b$ permit good prediction of the deformation of an ideal square-wave input as it travels from the point of insertion in the system to the point of detection.

### 11.5.2. Kinetics in Single-Channel Unsegmented Sample–Reagent(s) Processing Systems

The kinetics of mass transport (as discussed earlier in this chapter) and the chemical kinetics of continuously moving streams dictate the time distribution of chemical species in the plug sensed in these systems. Unsegmented continuous-flow systems have had prior developments in process control methodology and in studies of blood transport in the cardiovascular system. As early as 1970, Harris and Newman (40), discussing indicator dilution curves in physiological studies, recognized that "the theory and experiments in physical systems have shown that mixing patterns, interphase transport, and chemical reaction may have a marked effect on the shape of indicator curves."

The complete kinetic picture in these systems, which is still under development, then must consider axial and radial diffusion, convection, and the dynamics of chemical reactions.

Since laminar flow predominates in the transport of matter along tubes (8), although it may be disturbed at different points in the overall system (e.g., point of sample introduction, connecting parts, and detection zone), the starting point for most descriptions of physical kinetics is physical dispersion under laminar-flow conditions. These can be described by

$$\frac{dC}{dt} + U_{\max}\left[1 - \frac{r^2}{a^2}\right]\frac{dC}{dx} = D\left[\frac{d^2C}{dx^2} + \frac{dC}{r\,dr} + \frac{d^2C}{dr^2}\right] \qquad (11.27)$$

where $C =$ the concentration of dispersed species in molar units; $t =$ the time in seconds; $U_{\max} =$ the maximum linear velocity on the center axis of the tubing (in centimeters per second); $r =$ the radial distance from the center axis of the tubing (in centimeters); $x =$ the axial distance from the injection point (in centimeters); $D =$ the molecular diffusion coefficient (in centimeters squared per second); and $a =$ the radius of the tubing (in centimeters). This equation is transformed to a dimensionless form because that simplifies mathematical operations and because it is convenient to have flow parameters including the overall effect of reactor dimensions and flow mechanics:

$$\frac{dC^*}{d\tau} + \left[\frac{1-y^2}{\tau^{1/2}} - \frac{\Omega}{2\tau}\right]\frac{dC^*}{d\Omega} = \frac{1}{\tau N_{Pe}^2}\frac{d^2C^*}{d\Omega^2} + \frac{1}{y}\frac{dC^*}{dy} + \frac{d^2C^*}{dy^2} \qquad (11.28)$$

in which

$$\tau = \frac{D(t)}{a^2} \qquad \Omega = \frac{x(D/t)^{1/2}}{a(U_{max})}$$

$$y = \frac{r}{a} \qquad C^* = \frac{C}{C_0} \qquad N_{Pe} = \frac{a(U_{max})}{D} \qquad \text{(Peclet number)}$$

Ananthakrishnan et al. (41) numerically solved this equation by the method of "alternating direction implicit finite difference approximation" (42) for a single sample–carrier boundary resulting from a constant flow injection. Vanderslice et al. (43), in a significant contribution to the description of dispersion in laminar-flow injection systems, adapted the approach to solve for the dispersion occurring at both boundaries of an injected sample plug.

The role that chemical kinetics may have on the concentration–time distribution was documented in 1981 by Painton and Mottola (44), and a first-order chemical reaction term was integrated, in a cyclic form, to eq. 11.28 by the same authors (45). It is of interest to single out two observations from these studies: (a) The natural logarithm of the residence time decreases linearly with an increase in rate coefficient ("dilutions" resulting from chemical and physical contributions affect each other). (b) The rate coefficient as a function of the reduced time exhibits a periodic form (variations of rate coefficient along the plug are interpreted as the result of a periodic change in reaction order). The cyclic fluctuations of rate within the reacting plug can be located in leading, central, and trailing portions of the plug. In the leading and trailing portions the carrier–sample plug interfaces induce molecular diffusion, while the velocity profile (predominantly laminar) increases convection. The lack of interface in the central zone means that convection is the prevalent transport (equalization) process. Further application of this approach, under conditions as close as possible to laminar, was presented by Wada et al. (46).

A different approach to the description of dispersion in unsegmented (predominantly laminar) flow systems for mixing and reacting chemical species was taken by Betteridge et al. (47). By computer simulated physical and chemical contributions to dispersion in a single-channel flow injection system by a random-walk (stochastic, Markovian chain) method. The model concentrates on the movement of individual molecules and uses $10^3$ model molecules to represent the fate of $10^{23}$ molecular units in a conceptually sound approach. No comparison with actual experimentally obtained signals is included in this study, but the general conclusions extracted from the simulation are in agreement with published experimental findings. The model shows, for instance, that the molecules accumulating in the leading portion of the plug are the last to undergo reaction, as was experimentally demonstrated (44).

One of the earliest models applied to describe unsegmented continuous-flow systems is the tanks-in-series model (48). The basic concepts are similar to those underlying the plate theory of chromatography. The fundamental assumption is that the plug passes in a sequential manner through a number of idealized compartments (tanks) in which instantaneous mixing occurs. The model gives adequate description when the number of tanks is relatively large. This situation can be realized when mixing is aided by use of the single-bead-string reactor (SBSR) (see Chapter 9), and Reijn et al. (49) have successfully applied the model in such a case. The SBSR provides adequate mixing and keeps axial dispersion low. The slow complex formation of Cr(III) with EDTA was used as the reaction in a model reactor comprised of 280 imaginary tanks. These studies show that by using a SBSR with sufficiently large number of tanks, reaction rates (chemical kinetics) do not play a significant role in dictating dispersion.

### 11.5.3. Kinetics in Unsegmented Continuous-Flow Systems Using Mixing Chambers

Mottola and Hanna (50) adopted a series reaction (processes) model to describe peaklike signals obtained by injecting a sample directly into a well-stirred detection zone in an unsegmented continuous-flow system with re-circulation. The general scheme assumes the transient signal to be the result of simultaneous operation of two consecutive chemical reactions,

$$(A) + B \xrightarrow{\ k_1\ } C \tag{11.29}$$

$$C + (X) \xrightarrow{\ k_2\ } P \tag{11.30}$$

or a fast chemical reaction followed by an instrumentally imposed process (e.g., flow) in place of reaction 11.30, such as

$$(A) + B \xrightarrow{\ k_1\ } C_D \tag{11.31}$$

$$C_D \xrightarrow{\ k_2'\ } C' \tag{11.32}$$

In all cases the rate expression for process 11.32 is first-order in intermediate C, and in most cases the rate expressions for reactions 11.29 to 11.31 involve a single concentration variable, since (A) and (X) represent reagents present in sufficiently large concentrations that the corresponding reactions are pseudo-zero-order with respect to them. In some cases reactant X can be absent and the process of interest is merely the decay of a transient intermediate. Species $C_D$ in reaction 11.31 and process 11.32 represents species C in the detection area, whereas species C' is the same species after passing through the detector.

Constants $k_1$, $k_2$, and $k_2'$ represent rate coefficients characterizing the first-order or pseudo-first-order processes taking place in the mixing (detection) chamber. The individual values and ratios of these rate coefficients largely determine parameters of analytical significance. They control sensitivity, for instance, since the peak value of transient signal $S$ is

$$S_{max} = [B]_0 \kappa^{\kappa/(1-\kappa)} \tag{11.33}$$

where $\kappa = k_2/k_1$ or $k_2'/k_1$. As $\kappa$ decreases ($k_1$ becoming larger than $k_2$), $S_{max}$ approaches the limiting theoretical value of the signal in absence of reaction 11.30 or process 11.32. A value of $k_1$ of about five times the value of $k_2$ gives signal heights representing recoveries over 60% of the theoretical value. From a practical viewpoint little is gained by increasing the $k_1/k_2$ ratio above 100 (corresponding to about 95% signal recovery); for a 100% recovery $k_1$ must be more than $10^4$ times $k_2$ (50). In cases where sensitivity may be sacrificed, even values of $k_1$ slightly less than or about twice $k_2$ may be satisfactory. However, larger values of $k_1$ are desirable not only to enhance signal recovery but to shorten the time to reach $S_{max}$, since

$$t_{max} = (\ln k_2 - \ln k_1)/(k_2 - k_1) \tag{11.34}$$

Values of $k_1/k_2$ of 10 or more assure a short time to attain $t_{max}$. This ratio and a relatively large value for $k_2$ also assure a short time for return to baseline. Consequently the values of $k_1$ and $k_2$ control both sensitivity and the number of samples that can be injected per unit time.

Often the values of $k_1$ and $k_2$ can be determined separately, but in other cases the isolation of such consecutive processes is difficult. Direct estimation of $k_1$ and $k_2$ from the transient signal is possible (50) and of help in method development (50, 51).

Pungor et al. (52) mathematically described the signal obtained in a single-channel system when a miniature mixing chamber was located very close to or in the detection zone itself. The time dependence of the concentration of a given chemical species in the mixing chamber is described by

$$d(\Delta C_t)/dt = (F/V)\,[\Delta(C_s)_t - \Delta C_t] \tag{11.35}$$

where $F$ = the volumetric flow rate (mL/min), $V$ = the volume of the mixing chamber, $C_t$ = the actual analyte concentration at time $t$, $(C_s)_t$ = the concentration difference between carrier stream and the plug carrying the injected sample, and $\Delta$ refers to values after injection. The work of Pungor et al. is focused on electrochemical detection; for example, peak signal measurement in potentiometric detection is given by

$$\Delta E_{\max} \simeq \xi \log(k/C_0) + \xi \log M \tag{11.36}$$

where $\Delta E_{\max}$ = peak potential, $M$ = the mass of the sample injected, $k = \Delta C_{\max}/M$ = constant that is a function of the geometry of the experimental setup and flow conditions, $\Delta C_{\max}$ = the concentration at peak maximum, $C_0$ = analyte concentration in the carrier stream before injection of solution, and $S$ = the slope of the Nernst-type plot under equilibrium conditions. The generalized mathematical description of the model can be given by

$$\frac{d\Delta C}{dt} = \frac{F}{V}\left\{\frac{M}{2A'(\pi D_{\mathrm{eff}}\,t)^{\frac{1}{2}}}\exp-\frac{1}{4D_{\mathrm{eff}}}\left[\frac{L}{t^{1/2}}-\left(\frac{Ft^{1/2}}{A'}\right)^2\right]\right\} \tag{11.37}$$

where $A'$ = the cross-sectional area of the tubing (in cm$^2$), $D_{\mathrm{eff}}$ = the effective diffusion coefficient (which is not a simple physical constant but a composite parameter including several processes), and $L$ = the length of the tube (in centimeters) in which injected and carrier solution interact. $D_{\mathrm{eff}}$ depends on the radius of the transporting tubing, the flow rate, and the properties of the injected and carrier solutions.

Pardue and Fields (53, 54) proposed mathematical expressions to describe (kinetically) the process of dilution with or without chemical reaction in a mixing chamber crossed by a stream (with and without reagent) into which the sample is introduced. Stirring takes place in the mixing chamber. The kinetic description centers on a variable-time approach mathematically aided by the tank-in-series model applied to a single tank situation. Although the tank-in-series model is of questionable value for a rather small number of tanks, success in the case of a single tank (mixing chamber) under well-stirred conditions is not surprising. Basic assumptions in the model development are that (a) fast chemical reactions are involved (i.e., chemical kinetics is not limiting); (b) the sample is introduced in "plug flow" fashion; (c) dispersion from the point of sample introduction to the point of entrance in the mixing chamber is negligible and can be ignored (short distance between point of injection and chamber); (d) "instantaneous" mixing takes place in the well-stirred chamber; (e) dispersion is also negligible in the part of the system comprised of the chamber and the detector, and (f) the process can be physically divided in time steps as illustrated in Figure 11.5. Consider the hypothetical system

$$A + zR \rightarrow P$$

where $A$ = the chemical species of interest in the sample, $R$ = the reagent, and $P$ = the products. The time interval $t_1$ needed for all reagent originally filling the chamber to be displaced from the gradient chamber is given by

$$t_1 = (F/V) \ln\{1 + z([R]_{0,\,ch}/[A]_s)\} \tag{11.38}$$

where $[R]_{0,\,ch}$ is the reagent concentration in the chamber when $A$ first begins to enter the chamber, and $[A]_s$ is the concentration of $A$ in the sample.

Between $t_1$ and $t_2$, the concentration of $A$ in the chamber increases according to

$$[A]_{ch} = [A]_s\{1 - \exp[-(F/V)(t - t_1)]\} \tag{11.39}$$

where $t_1 \leq t \leq t_2$. The value of $[A]$ reaches a maximum at $t_2 = V_s/F$ ($V_s$ = sample volume introduced in the system), which is given by

$$[A]_{ch}^{max} = [A]_s\{1 - \exp[-(F/V)(t_2 - t_1)]\} \tag{11.40}$$

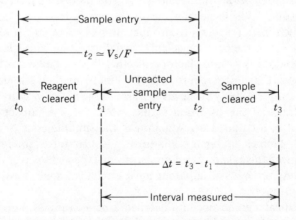

**Figure 11.5.** Time-steps physical model for a single-channel flow injection system with mixing chamber. $V_s$ = sample volume introduced into the system, and $F$ = volumetric flow rate. [From Ref. 53. Reproduced with permission of Elsevier Scientific Publishing Co.]

After $t_2$ the concentration of A in the chamber decreases according to

$$[A]_{ch} = [A]_{ch}^{max} \exp\left[-(F/V)(t-t_2)\right]$$
$$-z[R]_0\{1-\exp\left[-(F/V)(t-t_2)\right]\} \qquad (11.41)$$

where $t_2 \le t \le t_3$, and $[R]_0$ = the reagent concentration in the flow stream. The first term on the right-hand side of eq. 11.41 accounts for mass transport when A and R do not chemically interact. The second term takes into account the chemical interaction between A and R. A pure exponential decay results only when $z[R]_0 = 0$.

The main limitation of the treatment (54) is the assumption of plug flow. The extent of deviation from ideal behavior depends mainly on concentration values. Negligible deviations can be expected for low concentrations (0 to 50 mM); 10% or larger deviations can be expected for concentrations of 100 mM or larger. Details of the Pardue and Fields treatment can be found in the original publications (53, 54) and in pertinent monographs (55).

## 11.6.  OTHER KINETIC ASPECTS OF RELEVANCE IN ANALYTICAL CHEMISTRY

The examples dealt with in some detail in previous sections illustrate the intimate role played by kinetic concepts in well-established areas of analytical interest. Table 1.5 pointed out some additional areas. An obvious area has not been mentioned until now: the kinetic and mechanistic aspects of chemical reactions per se. It is obvious that studies directed to collect kinetic information and describe mechanistically the process(es) by which a chemical system evolves from reactants to products in its progress to equilibrium are relevant. However, their relevance may need illustration with a few examples.

In a review of the various factors affecting kinetic-based methods, Mark (56) points out that "Any report of a new kinetic method should include, if possible, a complete study of the mechanism of the reaction(s) involved . . . ." The importance of this assertion is illustrated with the impact on method development of a detailed investigation of the simultaneous (in situ) determination of amines in binary mixtures (57). Careful examination of the reaction(s) mechanism(s) revealed numerous unexpected sources of error that explained why, in spite of apparent ideal analytical conditions, quite poor results were obtained.

The analytical significance of a detailed kinetic study of a given catalytic system can be found in the work of Rodriguez and Pardue with the Sandell–Kolthoff reaction (58). Results of detailed studies of the osmium (59) and iodine (60) catalysis of the Ce(IV)–As(III) reaction are utilized for simultaneous determinations of the catalysts without prior separation. Iodide and osmium are simultaneously determined taking advantage of the Hg(II) inhibition of the iodide catalysis. Iodine is an effective catalyst, but iodate shows no catalytic action. Arsenic(III), however, slowly converts iodate to a form that has catalytic activity equal to an equivalent amount of iodide. These behaviors were exploited for the determination of iodide and iodate in mixtures with $[I^-]/[IO_3^-]$ ratios of 10:1 and 1:10 and the lower component present at $2.7 \times 10^{-8}\ M$.

The differentiation between a promoting and a true catalytic effect (see Chapter 4) as revealed by a kinetic study can explain why a catalytic like method fails under given conditions. A catalytic like determination of immobilized aldehyde groups was proposed by Shapilov (61) based on the rate-accelerating effect of aldehydes in the $H_2O_2$ oxidation of $p$-phenylenediamine (62). Attempts to utilize Shapilov's so-called catalytic determination resulted in contradictory behavior in a continuous-flow–stopped-flow system employed to determine immobilized aldehyde groups on the walls of open-tube enzyme reactors for continuous-flow determinations. The catalytic response was observed to decrease with repetitive determinations using the same reactor with attached glutaraldehyde, and the response for the same reactor showed a decreasing "catalytic" activity from day to day. A re-examination of the effect of glutaraldehyde demonstrated that the aldehyde acts as a promoter and not as a true catalyst (63) and that the product of the reaction is destroyed by further reaction with $H_2O_2$.

Although many more examples could be cited, those mentioned here are sufficient to point to the practical insights to be gained by acquiring basic kinetic information and understanding mechanistic considerations. Details of the role of kinetics in other analytical areas (e.g., solvent extraction and thermal methods) could equally well be added to this chapter, but the cases treated in some detail here should be sufficient to illustrate the point.

## REFERENCES

1. H. A. Mottola, *J. Chem. Educ.*, **58**, 399 (1981).
2. B. Kratochvil, D. Wallace, and J. K. Taylor, *Anal. Chem.*, **56**, 113R (1984).
3. H. L. Pardue, *Quim. Anal. (Barcelona), II (Extra)*, 24 (1983).
4. H. A. Mottola, *Anal. Sci.*, **2**, 317 (1986).
5. W. C. Gardiner, Jr., *Rates and Mechanisms of Chemical Reactions*, Benjamin, New York, 1969, pp. 165–170.
6. B. L. Karger, L. R. Snyder, and C. Horvath, *An Introduction to Separation Science*, Wiley, New York, 1973, Chapter 3.

7. A. J. Bard and L. R. Faulkner, *Electrochemical Methods: Fundamentals and Applications*, Wiley, 1980, pp. 127–134.

8. M. Valcarcel and M. D. Luque de Castro, *Flow Injection Analysis: Principles and Applications*, Horwood, Chichester, U.K., 1987, Chapter 3.

9. M. K. Kemp, *Physical Chemistry: A Step-by-Step Approach*, Dekker, 1979, pp. 67–69.

10. A. J. P. Martin and R. L. M. Synge, *Biochem. J.*, **35**, 1358 (1941).

11. J. J. Van Deemter, F. J. Zuiderweg, and A. Klinkenberg, *Chem. Eng. Sci.*, **5**, 271 (1956).

12. J. C. Giddings, *Dynamics of Chromatography*, Part 1; *Principles and Theory*, Dekker, New York, 1965.

13. J. C. Giddings, *J. Chem. Educ.*, **44**, 704 (1967).

14. P. H. Rieger, *Electrochemistry*, Prentice-Hall, Englewood Cliffs, N.J., 1987, Chapter 5.

15. W. E. Geiger and M. D. Hawley, in *Physical Methods of Chemistry*, Vol. 2: *Electrochemical Methods*, 2nd ed., B. W. Rossiter and J. F. Hamilton, Eds., Wiley, New York, 1986, pp. 2–3.

16. B. V. L'vov, *Inzh. Fiz. Zh.*, **2**, 44 (1959).

17. B. V. L'vov, *Spectrochim. Acta*, **17**, 761 (1961).

18. A. Syty, *CRC Crit. Rev. Anal. Chem.*, **4**, 155 (1974).

19. D. R. Thomerson and K. C. Thompson, *Chem. Br.*, **11**, 316 (1975).

20. C. W. Fuller, *Analyst (London)*, **99**, 739 (1974); **100**, 229 (1975); **101**,798 (1976); *Proc. Anal. Div. Chem. Soc.*, **13**, 273 (1976).

21. K. J. Laidler, *Chemical Kinetics*, 3rd ed., Harper & Row, New York, 1987, pp. 279–285.

22. J. A. Holcombe, R. H. Eklund, and J. E. Smith, *Anal. Chem.*, **51**, 1205 (1979).

23. G. Muller-Vogt and W. Wendl, *Anal. Chem.*, **53**, 651 (1981).

24. B. V. L'vov and P. A. Bayunov, *Zh. Anal. Khim.*, **36**, 837 (1981); *J. Anal. Chem. USSR (Engl. Transl.)*, **36**, 561 (1981).

25. B. V. L'vov, P. A. Bayunov, and G. N. Ryabchuk, *Zh. Anal. Khim.*, **36**, 1877 (1981); *J. Anal. Chem. USSR (Engl. Transl.)*, **36**, 1313 (1981).

26. J. Musil and I. Rubeska, *Analyst (London)*, **107**, 588 (1982).

27. B. V. L'vov and A. S. Savin, *Zh. Anal. Khim*, **38**, 1924 (1983); *J. Anal. Chem. USSR (Engl. Transl.)*, **38**, 1475 (1983).

28. B. V. L'vov and A, S. Savin, *Zh, Anal. Khim.*, **38**, 1933 (1983); *J. Anal. Chem. USSR (Engl. Transl.)*, **38**, 1481 (1983).

29. C. Th. J. Alkemade and P. J. Th. Zeegers, "Excitation and De-Excitation Process in Flames," in *Spectrochemical Methods of Analysis*, J, D. Winefordner, Ed., Vol. 9 of *Advances in Analytical Chemistry and Instrumentation*, C. N. Reilley and F. W. McLafferty, Eds., Wiley-Interscience, New York, 1971, pp. 22–34.

30. K.-P. Li, *Anal. Chem.*, **53**, 317 (1981).

31. K.-P. Li, *Anal. Chem.*, **53**, 2217 (1981).

32. H. A. Mottola, *Anal. Chem.*, **53**, 1312A (1981).

33. C. C. Painton and H. A. Mottola, *Anal. Chim, Acta*, **154**, 1 (1983).

34. G. Horvai and E. Pungor, *CRC Crit. Rev. Anal. Chem.*, **17**, 231 (1987).

35. R. E. Thiers, R. R. Cole, and W. J. Kirsch, *Clin. Chem. (Winston–Salem, N.C.)*, **13**, 451 (1967).

36. W. H. C. Walker, in *Continuous-Flow Analysis: Theory and Practice*, W. B. Furman, Ed., Dekker, New York, 1976, Chapter 7.

37. W. H. C. Walker, *Clin. Chim. Acta*, **32**, 305 (1971).

38. W. H. C. Walker, J. C. Shepherdson, and G. K. McGowan, *Clin. Chim. Acta*, **35**, 455 (1971).

39. P. L. Alger, *Mathematics for Science and Engineering*, McGraw-Hill, New York, 1957, Chapter 13.

40. T. R. Harris and E. V. Newman, *J. Appl. Physiol.*, **28**, 840 (1970).

41. V. Ananthakrishnan, W. N. Gill, and A. J. Barduhn, *AIChE J.*, **11**, 1063 (1965).

42. B. Carnahan, H. A. Luther, and O. J. Wilkes, *Applied Numerical Methods*, Wiley, New York, 1969, pp. 270–272.

43. J. T. Vanderslice, K. K. Stewart, A. G. Rosenfeld, and D. J. Higgs, *Talanta*, **28**, 11 (1981).

44. C. C. Painton and H. A. Mottola, *Anal. Chem.*, **53**, 1713 (1981).

45. C. C. Painton and H. A. Mottola, *Anal. Chim. Acta*, **158**, 67 (1984).

46. H. Wada, S. Hiraoka, A. Yuchi, and G. Nakagawa, *Anal. Chim. Acta*, **179**, 181 (1986).

47. D. Betteridge, C. Z. Marczewski, and A. P. Wade, *Anal. Chim. Acta*, **165**, 227 (1984).

48. O. Levenspiel, *Chemical Reaction Engineering*, 2nd ed., Wiley, New York,,1972.

49. J. M. Reijn, H. Pope, and W. E. Van der Linden, *Anal. Chem.*, **56**, 943 (1984).

50. H. A. Mottola and A. Hanna, *Anal. Chim. Acta*, **100**, 167 (1978).

51. S. M. Ramasamy, M. S. A. Jabbar, and H. A. Mottola, *Anal. Chem.*, **52**, 2026 (1980).

52. E. Pungor, Z. Feher, G. Nagy, K. Toth, G. Horvai, and M. Gratzl, *Anal. Chim. Acta*, **109**, 1 (1979).

53. H. L. Pardue and B. Fields, *Anal. Chim. Acta*, **124**, 39 (1981).

54. H. L. Pardue and B. Fields, *Anal. Chim. Acta*, **124**, 65 (1981).

55. M. Valcarcel and M. D. Luque de Castro, *Flow Injection Analysis: Principles and Applications*, Horwood, Chichester, U.K., 1987, Chapter 8.

56. H. B. Mark, Jr., *Talanta*, **20**, 257 (1973).

57. R. A. Grienke and H. B. Mark, Jr., *Anal. Chem.*, **38**, 1001 (1966).

58. P. A. Rodriguez and H. L. Pardue, *Anal. Chem.*, **41**, 1376 (1969).

59. R. L. Habig, H. L. Pardue, and J. B. Worthington, *Anal. Chem.*, **39**, 600 (1967).

60. P. A. Rodriguez and Pardue, *Anal. Chem.*, **41**, 1369 (1969).

61. O. D. Shapilov, *Zh. Anal. Khim.*, **35**, 2199 (1980); *J. Anal. Chem. USSR (Engl. Transl.)*, **35**, 1429 (1980).

62. G. Woker, *Ber. Dtsch. Chem. Ges.*, **47**, 1024 (1914).

63. J. C. Thompsen and H. A. Mottola, *Anal. Chem.*, **56**, 2834 (1984).

# EPILOGUE

The acquisition of knowledge is an ongoing process, and practices in analytical chemistry will continuously change. What will not change are the roots of analytical methodology, because the two pillars of the physical sciences will continue to be dynamical and steady-state components. The enduring fact that all chemical and physical processes are dynamically driven toward equilibrium cannot be endangered by any advances in high technology or modern science. Understanding of the steps involved will be perfected, and the dynamic routes to a steady state will be better understood and utilized in the future. This book, like practically all other books, should not be viewed as an end point.

My intention has been to focus on rate-based methods as they are practiced today and on some kinetic issues relevant to contemporary analytical chemistry that in my opinion have not been addressed sufficiently in the past.

Kinetics will have to be increasingly recognized as a basic component in analytical chemistry. This conviction was the motivation for writing this monograph.

# INDEX

Absorbance, 42, 118, 189, 245
  derivative circuit, 190
  measurement by multipoint methods, 221,
    225
Absorbance-time profiles, for Cu atomiza-
    tion, 246
Absorption-stat, 186
  control unit, 188
  setup, 187
Acceptor, in induced reactions, 80
Accuracy:
  in catalytic end-point indication, 74
  in pseudo-first-order method of propor-
    tional equations, 136
Acetaminophen, determination in phar-
    maceuticals, 119
Acetic anhydride, 76, 133
Acetoacetate, interference in Jaffé method,
    222
Acetylcholine, 68
Acetylcholinesterase, 68
N-Acetylcysteine, determination by catalytic
    currents, 95
Acid(s):
  acetic, 25, 75
  aminopolycarboxylic, 76
  butyric, 77
  citric, promotion, 78, 80
  $\beta$-$D$-gluconic, 67, 104
  lactic, 66
  malonic, 80, 81
  nitrilotriacetic, 76
  organic, 25
  oxalic, oxidation, 39
  oxalic, promotion, 78, 80
  perchloric, 75
  pyruvic, 66
  succinic, 80
Acid-base reactions, in catalytic end-point in-
    dication, 76
Acridine, 130
Activation, 7, 24, 44, 72
  applications, 76
  in catalytic end-point indication, 76
  enzymatic, 60, 81
Activation energy, in ionic evaporation, 247
Activator(s):
  allosteric, 81

definition, 76
Active sites, in enzymes, 58
Activity, 58
  of enzyme catalytic centers, 60
  factors affecting enzyme, 60
  molecular, 60
  retention in enzyme immobilization, 101
  specific, 60
Adenosine-5'-triphosphate, determination
    by chemiluminescence, 165
Adrenaline, determination, 95
Adsorption, in electrochemical processes, 242
Air, effect in simultaneous determination of
    sulfur compounds, 141, 142
Air-segmented continuous-flow:
  block diagram, 179
  kinetics, 249
  mixing by, 178
  sample dispersion in, 180
Alanine, determination:
  by chemiluminescence, 163
  of flavin adenine dinucleotide with, 165
  by uncatalyzed reaction, 118
Albumin, serum, determination by chemi-
    luminescence, 166
Alcohol(s), determination in blood, 64
  in mixtures, 133, 137
Aldehyde groups, determination of immo-
    bilized, 259
Aldrin, determination, 85
Alkaline phosphatase:
  inhibition, 85
  use in zinc determination, 82, 83
Alternating-direction implicit finite dif-
    ference approximation, use to describe
    kinetics in continuous flow, 253
Alumina, as support in phosphorimetry, 154
Aluminum, determination:
  by differential rate method, 125
  with fluorescence monitoring, 152
  in mixture with its citrate complex, 142
  by rate uncatalyzed reaction, 115
Amines, determination:
  aliphatic, by rate of uncatalyzed reaction,
    118
  by differential rates, and mechanistic im-
    plications, 258
  di- and poly-, by chemiluminescence, 166

265

Amines, determination (*Continued*)
  method of Roberts and Regan, 136
  tertiary, 25, 76
apo-*D*-Amino acid oxidase, in flavin adenine
    dinucleotide determination, 165
Amino acids:
  determination:
    by chemiluminescence, 164
    by differential rates, 143
    D- and L-, 68, 93
    by inhibition, 73
    by uncatalyzed reaction, 118
  in enzyme immobilization, 100
  in enzymes, 54
Amino groups, primary, role in enzyme im-
    mobilization, 103
1-Amino-4-hydroxyanthraquinone, in van-
    adium determination, 117, 152
Aminopeptidase, pig kidney, in zinc deter-
    mination, 82
Aminosilanes, in enzyme immobilization, 103
Ammonia, in pyridoxal 5'-phosphate deter-
    mination, 117
Ammonium hydrogen fluoride, glass etching
    with, 107
Ammonium ion sensing, in enzyme-catalyzed
    reactions, 68
Amperometric detection, in enzymatic deter-
    minations, 68, 93, 102, 104, 107
Amperometry, 93, 96, 189, 194
Ampicillin, determination, 131
Amplification, by enzyme cycling, 68, 69
Amylase:
  determination, 68
  immobilization, 100
Analog, 14
  reciprocal-time computer, 9
  system for catalytic end-point indication,
    203
  system for differential rate determinations,
    201, 202
Androsterone, determination, 161, 167
Aniline black, 2, 12
Antimony, electrocatalytic determination in
    effluent streams, 94
Antioxidants, determination in oil samples,
    120
Apoenzyme(s), 54, 82
  of tryptophanase, 117

Arsenazo III:
  as attacking ligand in thorium determina-
    tion, 117
  use in method of proportional equations,
    131
Arsenic (III):
  determination, 78
  induction effect, 80
Arsenic (III)/cerium (IV) reaction, *see*
    Cerium (IV)/arsenic (III) reaction
Arylamines, activation, 77
Asbestos, as support in phosphorimetry, 154
Ascorbic acid:
  determination, 119, 120
  determination by chemiluminescence, 164
  use in phosphorus determination, 116
Aspartic acid, determination, 164
Atomic absorption, nonflame, 21, 236, 247
  series processes, 222, 245
Automation, 170

Bacampicillin, determination by method of
    proportional equations, 131
Back-titration, in catalytic end-point indica-
    tion, 75
Bacterial luciferase, *see* Luciferase
1,2-Benzenediol (pyrocatechol), in vanadium
    determination, 2
Benzoic acids, substituted, determination by
    graphical differential rate method, 144
Benzophenone, determination by phos-
    phorimetry, 155, 158
Benzo[*a*]pyrene-7,8-dihydrodiol, determina-
    tion by chemiluminescence, 165
Betamethadone, determination by method of
    proportional equations, 132
Biacetyl (butanedione), as acceptor in phos-
    phorimetry, 155
Bile acids, determination by chemilumines-
    cence, 164
Bimolecular reactions:
  in differential reaction rate methods, 122
  in uncatalyzed determinations, 111
Bioluminescence, 158
  instrumentation, 193
Bipotentiometric monitoring, 197
2,2'-Bipyridine, activating effect, 78
Bis (2,4,5-trichlorophenyl) oxalate, in chemi-
    luminescence determinations, 164

Biuret:
  determination by fixed-time procedure, 120
  procedure for protein determination, 117
Blue tetrazolium reaction, in determination of corticosteroids, 119
Bolus flow, in segmented continuous-flow, 180
Bromate, 44
  in oscillating reactions, 81
Bromide, determination:
  by chemiluminescence in water samples, 163
  determination by method of proportional equations, 130
Bromine:
  determination by chemiluminescence in water samples, 163
  in the determination of organic species, 120
4-Bromobenzophenone, determination by phosphorimetry, 155
4-Bromobiphenyl:
  determination by phosphorimetry, 155
  phosphorescence spectrum, 156
Bromopyrogallol red, in silver determination, 44
Butylamine, determination in waste water, 118

Cadmium, determination in mixtures, 124, 143
Calcium ion:
  determination in mixture with magnesium, 137
  in enzyme activity, 81
  use in lipoprotein determination, 119
Calibration curve(s):
  for catalytic currents, 93
  for cobalt determination by chemiluminescence, 160
  in error analysis, 216, 217
  for method of Roberts and Regan, 136
  for single-point method, 126
Carbenecillin, differential rate determination, 125
Carbonic anhydrase, use:
  in DDT determination, 85
  in zinc determination, 82
apo-Carbonic anhydrase, use in zinc deter-

mination, 82
Carbonyl compounds, determination:
  by method of proportional equations, 136
  by method of Roberts and Regan, 136
Carboxymethylcellulose, in enzyme immobilization, 100
Carboxypeptidase, immobilization, 100
Carfecillin, determination by differential rate methods, 125
Carindacellin, determination by differential rate methods, 125
Catalase, 96
Catalysis:
  acid-base, 25
  in chemiluminescence, 159
  by cyanide ions, 43
  heterogeneous, 25, 88
  heterogeneous (enzymatic), 93
Catalyst, definition, 24
Catalytic currents, 91, 92, 95
  calibration graphs, 93
  determinations based on, 92
  voltammetric, 25, 91
Catalytic cycle, 24, 26, 70, 78, 91
Catalytic determinations, 1, 15, 24
  chemical reactions in, 36
  differential rate, in flow systems, 140
  mathematical basis, 25
Catalytic end-point determinations, 73–75
  and enthalpimetry, 198
Catalytic pre-wave, 92
Catalytic titrant, 73, 74
Catalyzed reactions:
  in biamperometric monitoring, 197
  modified, 72
Catechol, enzymatic determination in industrial samples, 65
CE, electrode process, 90
Cellophane, as support for enzyme immobilization, 161
Cellulosic supports, in phosphorimetry, 154
Centering the data, in differential reaction rate methods, 224
Centrifugal analyzer, sample disk, 188
Cerium:
  Ce(IV)/Ce(III) in oscillating reactions, 81
  determination by uncatalyzed reaction, 116
Cerium (IV)/arsenic (III) reaction, 2, 12, 36, 39, 40, 44–47

Cerium (IV)/arsenic(III)reaction (*Continued*)
  in differential rate determinations, 27, 140
  mechanistic implications, 254
Chemical kinetics:
  in continuous flow, 252, 253
  in flame spectroscopy, 246
  and the tanks-in-series model, 254
Chemical oxygen demand (COD), determination by uncatalyzed-reaction methods, 117
Chemical potential, and diffusion, 239
Chemical reactions, 15
  for catalytic determinations, 36
  coupled:
    classification, 90
    in heterogeneous catalysis, 89
    in optimization, 222
  in electrochemical processes, 242, 243
  fast, 16
Chemiluminescence, 158, 244
  in continuous-flow methods, 166
  detection, 189
  of $H_2O_2$-luminol reaction, 74
  instrumentation, 11, 193
Chloramines, in multipoint differential determination, 143
Chloramine T, in catalytic post-column detection, 49
Chlorate, oxidation by, 2, 12, 49
Chloride:
  effect on the Ce(IV)/As(III) reaction, 45
  determination by method of proportional equations, 130
Chlorine, differential determination, 143
Chlorine dioxide, gaseous, chemiluminescence determination in water samples, 162
2-Chlorobenzophenone, determination by phosphorimetry, 155
4-Chlorobiphenyl, phosphorescence spectrum, 156
Cholesterol, determination in blood, by chemiluminescence, 164
Cholesterol oxidase, in cholesterol determination, 164
Cholinesterase, inhibition in determination of organophosphorus compounds, 85
Choriogonadotropin, determination by chemiluminescence, in human urine, 165

Chromatogram, 240
Chromatography, 237
  diffusion in, 241, 242
  elution, 20
  kinetics in, 240
Chromium (III):
  in luminol chemiluminescence, 159
  promotion and, 78
  in tank-on-series model, 259
Chromium (VI), 78
  determination, 18, 44
    by catalytic current, 94
Cobalamine, in metalloenzymes, 82
Cobalt:
  in luminol chemiluminescence, 159
  in reactions on vitamin $B_{12}$, 82
Cobalt, determination:
  by catalytic currents, 94
  by electrogenerated chemiluminescence, 162
  by lucigenin chemiluminescence, 160
  in mixtures, 124, 125, 140
Coenzymes, 54
Cofactors, 54, 69, 81
Collisional quenching, 150
Computer(s), 14, 207. *See also* Microcomputer(s)
  in automation, 171
  bipolar pulse conductometry, control by, 197
  data manipulation by, 237
  and method of proportional equations, 128, 139
  mini-, and multipoint differential determinations, 143
  processing of multipoint data, 224
  reciprocal time, 201
  simulation:
    of catalytic end-point, 76
    of chemical reactions, 207
    of dispersion in continuous-flow, 253
  in stopped-flow mixing, 227
Concentration, relative error, in catalytic determinations, 42
Conductance, measurement, 197
  by bipolar pulse, 11, 197
Consecutive processes, 222
  in furnace atomic absorption, 245
  in unsegmented continuous-flow mixing, 254

Continuous flow, 178
  absorptiostat, 186, 187
  air-segmented, bile acid determination, 164
  chemiluminescence and, 166
  closed-loop, 96, 97
  determination of immobilized aldehyde
      groups, 259
  differential rate methods and, 137, 140,
      141
  electrochemical detection in, 197
  fixed-time measurements in, 182
  in monitoring of fast reactions, 178
  processing, 4, 8, 11, 18, 19, 21, 49, 68
  and stopped-flow, hybrid system, 175, 176
Controlled-pore glass, in enzyme immobiliza-
      tion, 107
Convection:
  in analytical chemistry, 238
  in continuous-flow, 183
  in electrochemical processes, 242
  in unsegmented continuous-flow, 252
Copper:
  acetate (Cu$^{2+}$), as reagent in differential
      rate determinations, 132
  activation by, 78
  catalysis, 37, 45, 49
  complexes, in inhibition, 73
  determination
    in biological samples, 19
    by chemiluminescence, 162
    in mixtures, 124, 143
    by uncatalyzed reaction, 116
  as ion scavenger in differential rate deter-
      minations, 137
  kinetic modeling, in atomic absorption,
      245
  in luminol chemiluminescence, 159
Correlation matrix, in differential rate
      methods, 224
Corticosteroids, determination:
  by chemiluminescence, 166
  in pharmaceuticals, 119
Coupling, of enzyme-catalyzed reactions, 67
Creatinine, determination:
  by multipoint curve-fitting, 222
  in serum by uncatalyzed reaction, 118
Creatinine phosphokinase (CPK), 62
Cresol purple, determination in mixtures, 131
Cresol red, determination in mixtures, 131
[2.2.1] Cryptand, as reagent for differential

rate determinations, 137
Crystallization, 20
Cumene, in determination of antioxidants,
      120
Cyanide ions:
  in heterogeneous catalysis, 89
  specific catalysis, 43
Cyanogen bromide, use in enzyme immobili-
      zation, 167
Cycling, of enzyme-catalyzed reactions,
      68, 69
Cyclotetramethylenenitramine, determina-
      tion in mixtures, 134
Cyclotrimethylenenitramine, determination
      in mixtures, 134
Cysteine, determination by catalytic cur-
      rents, 95

Data manipulation in analysis, 237
DCTA, see trans -1,2-Diaminocyclohex-
      anetetraacetate
DDT, see Dichlorodiphenyltrichloroethane
Dead time:
  in continuous-flow mixing, 178
  in stopped-flow mixing, 173
Debubbler, and signal deformation in air-
      segmented continuous-flow, 250
Decay periods, in fluorescence and phos-
      phorescence, 153
Decyl sulfate, in phosphorimetry, 155
Dehydrogenases, 54, 55
  glucose, 96–98
  glucose-6-phosphate, 69, 106
  glutamate, 69, 99
  3-α-hydroxysteroid, 164, 167
  isocitrate, 82
  lactate, 62, 66, 68, 107
  malate, 67
Demasking, in catalytic end-point indica-
      tion, 74
Denaturation, of enzymes, 60, 61, 69, 81
Density, mixing efficiency and, 180
Derivative (differential) methods, 29
  use with uncatalyzed reactions, 113
Derivative readout, in absorptiometric detec-
      tion, 189, 190
Desmolases, 55
Desorption, in electrochemical processes,
      242, 243
Detection, 189

Detection (*Continued*)
   in air-segmented continuous-flow, 179
   methods, chemiluminescence, 189
   methods, in enzyme-catalyzed reactions, 63
   thermochemical, 104
Detectors, electrochemical, 194
   bulk property, 196
   cascade, 196, 197
   porous-tubular, 196, 197
   thin-layer, 196, 197
   wall-jet, 196, 197
   wall-tubular, 196, 197
   wire type, 196, 197
Dialyzer, in air-segmented continuous flow,
      179
*trans*-1,2-Diaminocyclohexanetetraacetate:
   in differential reaction rate methods, 137
   in method of proportional equations, 131
   use in thorium determination by ligand ex-
      change, 117
1,4-Diamino-2,3-dihydroxyanthraquinone, in
      iron determination, 116
4,8-Diamino-1,5-dihydroxyanthra-
      quinone-2,6-disulfonate, in cerium
      and manganese determination, 116
*o*-Dianisidine (3,3'-dimethoxybenzidine),
      in coupling with enzyme reactions, 68
4,4'-Dibromobiphenyl, determination by
      phosphorimetry, 155
Dichlorodiphenyltrichloroethane (DDT), de-
      termination by enzyme inhibition, 85
2,6-Dichloroindophenol, in ascorbic acid de-
      termination, 119
2,6-Dichlorophenolindophenol, in kojic acid
      determination, 119
2,4-Dichlorophenoxyacetic acid (2,4-D), de-
      termination by enzyme inhibition, 85
Diethylenetriaminepentaacetate (DTPA), use
      in determinations by the method of
      proportional equations, 131
*N,N'*-Diethyl-*p*-phenylenediamine, use:
   in differential determination of chlorine
      and chloroamines, 143
   in single-point method, 127
Differential, phosphorimetry, 155, 158
Differential reaction rate methods, 4, 12,
      122, 125, 143, 258
   classification of, 123
   critical evaluation of, 137, 138
   instrumentation for, 201, 203

recommendations for, 224
regression computations in, 223
Diffusion, 19, 180
   in analytical chemistry, 238
   axial, 238, 252
   in chromatography, 241
   in immobilized enzyme systems, 104, 107
   in kinetics in flame spectroscopy, 248
   and mixing in unsegmented continuous
      flow, 184, 252
Diffusion coefficient(s):
   in chromatography, 240, 241
   effective, 256
   in Fick's laws, 239
   in unsegmented continuous-flow, 252, 253
Digital counting system, fixed-time measure-
      ments, 9, 199, 200
5,7-Dihalo-8-quinolinols, in differential reac-
      tion rate methods, 125
Dimethyl sulfide ($CH_3SCH_3$), determination
      by chemiluminescence, 162
*o*-Dinitrobenzene, in PCC determination,
      119
9,10-Diphenylanthracene, in chemilumines-
      cence determination of oxalate and
      urate, 163
2,2'-Dipyridyl ketone hydrazone, in catalytic
      determination of copper, 19
Dispersion:
   axial, and mixing efficiency, 180
   in flow systems, 252, 253
   in sample carryover in segmented contin-
      uous-flow, 180
   in unsegmented continuous-flow, 184
Disproportionation, in heterogeneous cataly-
      sis, 90, 92
Dissolved oxygen, monitoring in enzyme-
      catalyzed reactions, 96, 107
Double bonds (olefinic), single-point deter-
      mination, in synthetic rubbers, 4
Double-switching network:
   analog, 205
   hybrid, 206
Drift, in absorptiometric detection, 190
Dropping mercury electrode (DME), in cata-
      lytic currents, 94, 95
DTPA, *see* Diethylenetriaminepentaacetate
Dynamic range:
   in catalytic inhibition, 73, 74
   for $ClO_2$ determination, 62

in fluorescence detection, 152
for hydrogen peroxide determination, 107
in uncatalyzed determinations, 115
for vanadium determination, 152
Dysprosium, 131

$E_{1/2}$, 92
EC:
electrode process, 90, 91
numbers for enzymes, 62, 82
EC', electrode process, 90, 91
ECE, electrode process, 90, 92
Eddy movements:
in single-bead-string reactors, 183
turbulent, 238
EGTA, see Ethylene glycol bis(2-aminoethyl
ether)-$N,N,N',N'$-tetraacetic acid
Electrocatalysis, 88
Electrochemical, detection, 189, 194
Electrochemical detectors:
bulk property, 196
configurations, 196
surface type, 194
Electrochemical processes and diffusion,
242–244
Electrode:
dropping mercury, 94, 95
enzyme, 70. 100, 102
kinetics, 20, 242
rotating disk:
in catalytic currents, 94
in chemiluminescence, 162
Electrode reactions:
chemical and electrochemical steps in, 243
in heterogeneous catalysis, 15, 89, 90
Electronics:
ancillary, 198
in data manipulation, 237
Electron paramagnetic resonance 18
rapid scanning, 198
Electron transfer, activation and kinetics, 77,
243
Electrothermic recombination, in kinetics in
flame spectroscopy, 248
End-point indication, catalytic, 73, 76, 198
Enthalpimetry, 197
Enzymatic determinations:
with dissolved enzymes, 62, 93
of glucose, 64, 67, 93, 105
with immobilized enzymes, 99

Enzyme(s),
activation, 60, 81
active sites, 58
classification, 55
composition, 54
immobilized, 25, 70, 88, 100
recirculation, 96, 97
Enzyme activity, 53, 58
and error fluctuations, 216
retention in immobilization, 101
and stat-procedures, 186
units for, 58, 60
Enzyme-catalyzed reactions, 53
kinetics, 55
modification of, 81
monitoring, 63, 151
Enzyme concentration, 58
and method of proportional equations, 129
Enzyme denaturation, 60, 61, 69, 81
Enzyme–enzyme interaction, in enzyme immobilization, 100
Enzyme inhibition, 83
competitive, 83, 84
noncompetitive, 83, 94
uncompetitive, 83, 85
Enzyme reactor(s):
coimmobilized enzymes in, 106
in continuous-flow systems, 99
open-tubular-wall, 104, 106
packed, 104
single-bead-string, 107
Enzyme regeneration:
in homogeneous systems, 93
in immobilized systems, 99
Enzyme–substrate interactions, 83
Equilibrium:
measurements, comparison with kinetic
measurements, 103, 152, 153
predictive signal value at, 21
Erioglaucine, in heterogeneous catalytic
methods, 88
Error, 42
in uncatalyzed determinations, 113, 114
Error analysis, 212, 215–219, 221, 223–225
comparative studies in, 227
random, 214–216, 226
random, in two-rate measurements, 218
signal-to-noise ratio, 228
Ethanol, determination:
by activation of $Cu^{2+}$, 77

Ethanol, determination (*Continued*)
  by enzymatic method, 64
  by uncatalyzed rate, 119
Ethidium-DNA complex, 227
Ethyl acetate, single-point determination, 127
Ethylenediamine- *N,N,N',N'*-tetraacetic
    acid (EDTA), 73–76, 82
  in determinations with uncatalyzed reac-
      tions, 116
  differential methods and, 124
  in lipoprotein determination, 119
  in rare earths determination, 131
  tank-in-series model and, 254
Ethylene glycol bis(2-aminoethyl ether)-
    *N,N,N',N'*-tetraacetic acid (EGTA),
    76, 124
Excitation–deexcitation, 149
Excited state, 149
Exponential deformation, of signal in air-
    segmented continuous-flow, 250
Extrapolation:
  graphical, 138, 224
  graphical, analog system for, 202
  linear, 123, 144
  logarithmic, 12, 123, 124, 133, 134, 155,
      158

Faster-reacting component, in uncatalyzed
    determinations, 113
Fast optics, absorptiometric detection with,
    192
Fatty acid(s), oxidation of, 53
Feedback:
  control, 9
  optical, 9, 191
Fick's laws, 239
Firefly luciferase, *see* Luciferase
First-order processes, 17
  in atomic absorption spectroscopy, 245
  in chromatography, 241
  in error analysis, 223
  graphical extrapolation, 124
  irreversible, optimization, 222
  Kalman filtering in, 231
  kinetics and uncatalyzed determinations,
      112
  plots, 79
  rate coefficients, fluctuations, 213
  single-point method, critical evaluation,
      126

Fixed-time procedures, 2, 29, 30, 40, 41, 49
  in continuous-flow systems, 181
  for corticosteroids determination, 118
  for creatinine determination, 118
  digital counting systems for, 200
  error analysis, 215
  for lipoprotein determination, 119
  for PCC determination, 119
  for phenols determination, 117
  plotting in, 32
  rate computer for, 200, 227
  setup for continuous measurement, 182
  signal-to-noise in, 228
Flame spectroscopy, 246, 248
Flavin adenine dinucleotide, determination
    by chemiluminescence, 165
Flavin adenine mononucleotide oxidoreduc-
    tase, in NADH determination, 165
Flavin adenine mononucleotide(s):
  in bile acids determination, 165
  enhancement of chemiluminescence in SO$_2$
      determination, 162
Flow injection, using chemiluminescence, for
    determining,
  adenosine 5'-triphosphate, 165
  alpha-amino, 164
  copper, 162
  glucose, 163
  hydrogen peroxide, 162
  protein, 166
  serum albumin, 166
Flow injection analysis:
  manifold for, 182
  manifold for determination of glucose and
      sucrose, 105
Flow rate:
  effect on oxygen sensing, 96
  in mixing efficiency, 180
  secondary, 183
Flow-through systems, closed, 93
  diagram for, 97
Fluorescence, 18, 149, 151–153, 244
  in catalytic differential rate determinations
      in flow systems, 143
  detection, 151, 189, 193
  lifetime, and differential method, 125
  quenching, applications of, 130
Fluoride:
  effect on Ce(IV)-As(III) reaction, 46
  in unsegmented continuous flow, 181

2-Fluoroaldehyde 2-pyridylhydrazone, in determination of magnesium, 115, 152

Fluorometry:
determination of tryptophan, 119
monitoring in uncatalyzed reactions, 115, 116
reaction rate instrument, 11, 195
signal-to-noise ratio, 229

Flux, in kinetics of electrochemical processes, 243

Formaldehyde:
determination by uncatalyzed reaction, 118
use in tryptophan determination, 119, 153

Formazan, in corticosteroid determination, 119

Free energy, in light emission, 159

Fructose, determination in mixture with glucose by method of proportional equations, 131

Galactose, enzymatic determination, 64, 93

Gallium, determination:
in bauxite by uncatalyzed reaction rate method, 116
by differential rate method, 125

Gases:
determination by chemiluminescence, 161
leak detection by catalytic method, 48, 49

Gaussian distribution:
in air-segmented continuous-flow analysis, 250
in chromatography, 241
in least-square regression, 226

Germanium, determination as germanate, 116

Glass:
alkylamino-, in enzyme immobilization, 103
beads (porous), in enzyme immobilization, 163
controlled-pore-, in $H_2O_2$ determination by chemiluminescence, 162
in enzyme immobilization, 161
etching, 107

Glucose, determination:
by chemiluminescence, 163
enzymatic, 64, 67, 93
in soft drinks, 105
with fluorometric monitoring, 151

in mixture with fructose by method of proportional equations, 131

Glucose oxidase, 8, 55, 96, 99, 104
in glucose determination by chemiluminescence, 163

L-Glutamate, 67
determination by chemiluminescence, 163

Glutaraldehyde, in enzyme immobilization, 103, 107

Glutathione, determination by catalytic current, 95

Glycerol, and catalytic activation, 77

Glycine:
differential reaction rate determination, 143
as reagent in method of proportional equations, 131

Gold (III), in heterogeneous catalysis, 89

Graphical methods:
analog system for extrapolation, 202
critical evaluation of extrapolation, 138
and differential rate determinations, 123
extrapolation, error analysis, 224
interpolation, for three-component determination, 145, 146
linear extrapolation plot, 144
slope/intercept method, 143

Graphite:
furnace, kinetics in atomic absorption spectrometry, 245
rotating disk electrode, in catalytic currents, 94

Griess reaction, in determination of sulfonamides, 118

Hafnium, in selectivity of catalytic determinations, 44

Half-life (half-time), 17, 78
in continuous-flow mixing, 178
mixing of reactants and, 171
stopped-flow mixing and, 173

Halobiphenyls, determination by phosphorimetry, 155

Harman:
in fluorometric determination of bromide and chloride by method of proportional equations, 130

Heparin:
determination by chemiluminescence, 165
in determination of lipoprotein, 119

Heptachlor, determination by enzyme inhibition, 85

12-Heteropolymolybdates, use in the method of proportional equations, 132

Hexacyanoferrate:
determination, 78
in heterogeneous catalysis, 88
hydroxylamine oxidation by, 197
in induction, 80

Hexamethylenetetramine, determination in pharmaceutical preparations by uncatalyzed reaction, 118

Hexokinase, immobilized, for use in reactors, 106

High performance liquid chromatography (HPLC), post-column catalytic detection, 47

Histidine, differential rate determination, 143

Holmium, determination in mixtures, by method of proportional equations, 131

Holoenzymes, 54

Homocysteine, determination by catalytic currents, 95

Horseradish peroxidase, in determination of flavin adenine dinucleotide, 165

Hydrases, 55

Hydrogen:
activated, 88
bonding, in phosphorimetry, 154

Hydrogen peroxide, 8, 19, 37, 40, 44, 45, 67, 68, 96, 104
in aldehyde determination, 259
in chemiluminescence, 159
determination, 107
determination by chemiluminescence, 162
in determination of di- and polyamines, 166
differential rate determination in flow systems, 140
electrogeneration of, 159
in method of proportional equations, 132
in uncatalyzed determination of organic species, 120

Hydrogen sulfide:
and catalytic currents, 92
determination by chemiluminescence, 162

Hydrolases, 55, 63

Hydroquinone, determination, 120

2-Hydroxybenzaldehyde thiosemicarbazone:
in differential rate determinations in flow systems, 140
in method of proportional equations, 132

7-$\alpha$-Hydroxy bile acids, determination by chemiluminescence, 165

12-$\alpha$-Hydroxy bile acids, determination in human serum by chemiluminescence, 164

Hydroxy compounds, determination by method of Roberts and Regan, 136

Hydroxylamine hydrochloride, use in method of proportional equations, 136
oxidation, 197

2-Hydroxy-1-naphthaldehyde p-methoxybenzoylhydrazone, in determination of aluminum by uncatalyzed reaction rate method, 115

p-Hydroxyphenylacetate, in $H_2O_2$ determination, 107

8-Hydroxyquinoline:
in differential reaction rate methods, 125
in selectivity in catalytic methods, 44

8-Hydroxyquinoline-5-sulfonic acid, in determination of aluminum, 152

Hypochlorite, determination in drinking water by chemiluminescence, 160

Hypophosphite, 88, 89

Imidazole, use in determinations by method of proportional equations, 131

Immobilization of enzymes:
advantages, 99, 100
in chemiluminescence determinations, 161, 167
by covalent binding, 100, 104
definition, 99
electrostatic, 99
by membrane or gel entrapment, 99
by physical adsorption, 100
types of, 100

Immunoassay, for choriogonadotropin determination, 165

Indication, catalytic end point, 75

Indicator dilution curves, in kinetics in continuous-flow systems, 252

Indicator dilution technique, in unsegmented continuous-flow mixing, 181

Indicator reaction, 43
in catalytic end-point indication, 73

cerium (IV)-arsenic (III), 45
definition, 36
gallic acid-bromate, 49
iodide-BrO$_3$, 46
iodide-H$_2$O$_2$, 78
lucingenin-H$_2$O$_2$, 160
luminol-H$_2$O$_2$, 160
Malachite Green-periodate, 73, 74
1,10-phenanthroline-H$_2$O$_2$, 160, 162
Pyrocathecol Violet-H$_2$O$_2$, 73, 78
sulfanilic acid-persulfate, 78
sulfite–permanganate, 163
tris [1,10-phenanthroline]iron(II)-
Cr(VI), 79
Indium, determination by differential rate
method, 125
Induced reactions, induction factor, 80
Induction period(s), 32
in heterogeneous catalysis, 89
length, 34
Inelastic collision, in kinetics in flame spectroscopy, 248
Inhibition, 24, 46
applications, 73
enzymatic, 6
mixed (enzymatic), 85
partial or total, 73
of peroxidase activity, 107
Inhibitors, 25
definition, 73
in enzyme-catalyzed reactions, 83
in heterogeneous catalysis, 88
sulfur-containing, 88
Initial rate, 29, 40–42, 102
in absorbance detection, 189
conditions and enzyme catalysis, 55
in enzyme activity and activation, 61, 81
in error analysis, 215
fixed-time digital system, 199
in fluorescence determination of vanadium, 152
in inhibition, 84, 85
maximum, 56
in multipoint methods, 221
in pseudo-zero-order differential methods,
136
in uncatalyzed reactions, 112, 115, 118
Inorganic redox systems, 36
Integral methods, 29, 41
in uncatalyzed determinations, 113

Integrator:
for initial rate measurement, 199
for slope measurement, 198
Interferences, in kinetics in flame spectroscopy, 248
Internal conversion, 149
Intersystem crossing, 149, 150
Invertase, 100
immmobilized, 104, 105
Iodate, in Sandell–Kolthoff reactions, 45, 259
Iodide:
catalysis, 2, 40, 46, 259
catalysis of Ce(IV)-AS(III) reaction, 47
as catalyst in differential rate determination of chlorine and chloramines, 43
as catalytic titrant, 43
determination by method of proportional
equations, 130
determination simultaneously with thiocyanate, 140
in heterogeneous catalysis, 88
Iodinated thyronines, 49
Iodine:
catalysis and voltammetric currents, 94
determination in grass, sea water, serum,
urine, table salts, and vegetables, 45
in the determination of perbromate, 116
in Sandell–Kolthoff reactions, 258
in uncatalyzed determination of organic
species, 120
4-Iodobiphenyl, phosphorescence spectrum,
156
Ion exchange, 20
Ionic strength, effect on rate coefficients, 214
Iridium, fluorometric determination as catalyst, 151
Iron, determination:
by catalytic currents, 94
determination:
in mixtures by differential method, 124,
125
in mixtures by method of proportional
equations, 132
by differential reaction rate method in continuous-flow system, 140
in luminol chemiluminescence, 159
in mixtures by differential method, 124,
125
by uncatalyzed reaction rate method, 116
in uric acid determination, 119

Isobutylamine, determination in waste water, 118
Isokinetic contours, in graphical interpolation methods, 146
Isoleucine, determination:
  by differential rate method, 143
  by uncatalyzed reaction rate method, 118
Isopropyl acetate, determination by single-point method, 127

Jaffé reaction, in determination of creatinine, 118, 222
Jump, pressure or temperature, 18

Kalman filtering, 230
  algorithm for, 232
  extended, 230
Karl Fischer reagent, 160
Katal, 60
α-Ketoglutarate, in enzyme cycling, 131
Ketones, determination by pH-stat method, 202
Kinetics:
  in absorption/emission spectroscopy, 244
  in air-segmented continuous-flow, 249
  in analytical chemistry, 1, 236
  in chromatography, 240
  in continuous-flow systems, 248
  of coupled reactions, 222
  in electrochemistry, 242
  of enzyme-catalyzed reactions, 54
  in flameless atomization, 245
  in flame spectroscopy, 248
  in fluorescence, 149
  in unsegmented continuous-flow with mixing chambers, 254
Kojic acid, determination in fermentation media by uncatalyzed reaction rate method, 119

L -Lactate, determination, 107
  by chemiluminescence, 163
  by enzyme electrode, 70
Lactose, enzymatic determination, 65, 68
Lambert–Beer law, in catalytic determinations, 42
Laminar flow:
  in electrochemical detection, 197
  in unsegmented continuous-flow systems, 96, 183, 252

Landolt effect:
  reaction, 32
  reagent, 34
Lead, determination in mixtures by differential methods, 124
Least-squares regression, 224
  in Kalman filtering, 230
Leucine, determination by uncatalyzed reaction rate method, 118
Leuco crystal violet, in coupling of enzyme reactions, 68
Lifetime:
  fluorescence, 125
  of $S_1$ state, 150
Ligand exchange:
  in determination of copper by uncatalyzed reaction rate method, 116
  in determination of thorium, 117
  in logarithmic extrapolations, 124
  and method of proportional equations, 131, 132
  reactions, and determinations by uncatalyzed reaction rate methods, 115
Light emission:
  kinetic methods based on, 149
  source, stabilization, 190, 191
Limit of determination, 41
  in chemiluminescence determination of proteins, 160
Limits of detection:
  for alpha-amino acids, 164
  for aluminum by fluorescence determination, 152
  for bile acid determination by chemiluminescence, 164
  in catalytic determinations, 40–42
  for cholesterol determination by chemiluminescence, 164
  for cobalt determination by chemiluminescence, 160, 162
  for copper determination by chemiluminescence, 162
  in fluorescence determinations, 151
  for heparin determination by chemiluminescence, 165
  for $H_2O_2$ determination, 107
  for $H_2O_2$ determination by chemiluminescence, 162
  for iodide determination, 48
  for isomeric tetraiodothyronines, 49

for L-asparatic acid determination by chemiluminescence, 164
in lucigenin reaction, 160
for manganese determination in natural waters, 49
for metal ions in luminol chemiluminescence, 159
for NO determination by chemiluminescence, 161
for pyridoxal 5'-phosphate determination, 117
for serum albumin, 166
for $SO_2(g)$ determination by chemiluminescence, 163
for tetraiodothyronine and triiodothyronine, 47
for tryptophan determination, 153
in uncatalyzed reactions, 115
for uranium, in heterogeneous catalysis, 94
for zinc determination by enzyme reactivation, 82
Lindane, determination by enzyme inhibition, 85
Linear methods:
extrapolation, 123
graphical extrapolation plot, 144
least-squares and differential, 224
Lipase, 63
castor oil, pH dependence, 61
inhibition, 85
in lipoprotein determination, 118
pancreas, pH dependence, 61
Lipoprotein (serum low-density), determination by uncatalyzed reaction rate method, 118
Liquid leaks, detection by catalytic method, 48, 49
"Lock-key" analogy, in enzyme inhibition, 83
Luciferase:
bacterial:
immobilized, 161
in NADH determination, 165
firefly, in adenosine 5'-triphosphate determination, 165
firefly, in bile acids determination, 164
Lucigenin:
in chemiluminescence, 159
in heparin determination, 165
limits of detection of reaction, 160

Luminol:
chemiluminescence, 159
determination:
of $ClO_2$, 162
of di- and polyamines, 166
of flavin adenine dinucleotide, 165
of glucose, 163
oxidation, in bromine and bromide determination, 163
sonically induced chemiluminescence, 160
Lysine, differential reaction rate determination, 143

Magnesium, determination:
by differential reaction rate method, 125
by fluorometric rate measurement, 152
in mixture with calcium and strontium, 137
in plasma by enzyme activation, 82
by uncatalyzed reaction rate method, 115
L-Malate, determination by chemiluminescence, 163
Maltose, 64, 68
Manganese:
as catalyst, 39
as catalytic titrant, 73, 74
continuous-flow differential rate determination, 140
determination:
by activation, 75, 78
of dyes by method of proportional equations, 131
in fumes of industrial workplaces by uncatalyzed reaction, 116
in mixtures with iron by method of proportional equations, 132
in natural waters, 49
in isocitric dehydrogenase activation, 82
in luminol chemiluminescence, 159
Masking, in catalytic end-point indication, 74
Mass transfer:
in electrochemical processes, 242
radial, in air-segmented continuous-flow mixing, 180
Material balance, in chromatography, 240
Mean dwell time, and signal deformation, in air-segmented continuous-flow systems, 250
Mecillinam, epimers, determination by the method of proportional equations, 131

Mercury:
    differential rate determination, 143
    inhibition of iodide catalysis, 46, 259
    in thiamine determination, 118
Metal complex catalysis, 72
Metal-ion-catalyzed systems, modification
        of, 72
Metalloenzymes, use in modified-enzyme cat-
        alyzed reactions, 82
Methanethiol (methyl mercaptan $CH_3SH$),
        determination by chemiluminescence,
        162
4-Methylbenzophenone, determination by
        phosphorimetry, 155
3-Methylbenzothiazol-2-one hydrazone, use
        in method of proportional equations,
        132
Methyl mercaptan, see Methanethiol
Methyl parathion, determination by enzyme
        inhibition, 85
Micelle, stabilization in phosphorimetry, 155
Michaelis–Menten constant, 55, 57
    in enzyme immobilization, 101
    in enzyme inhibition, 84, 85
    in multipoint methods, 221
Microcomputer(s):
    in automation of on-line reagent dilution,
        176
    in stopped-flow mixing, 176, 224
Migration, in electrochemical processes, 242
Minimum detectable quantity:
    for $\alpha$-amino acids determination, 164
    for heparin determination, 165
    for iodide determination, 48
    for serum albumin, 166
Minimum determinable quantity, in chemi-
        luminescence determination of pro-
        teins, 160
Mixed ligand complexes, in activation, 77
Mixing:
    in air-segmented continuous-flow, 178, 180
    batch, 18
    in catalytic end-point indication, 184
    centrifugal, 187
    efficiency, 180
    by magnetic stirring, 184
    manual, 184
    by plunger type mixer, 172
    rapid, 18
    stirring button for, 184

stopped-flow, 172, 173
    in unsegmented continuous-flow, 183
Mixing jet, in stopped-flow mixing, 174
Molar absorptivity:
    in catalytic determinations, 42
    in error analysis, 221, 225
Molecular diffusion, in continuous-flow mix-
        ing, 180
Molybdate, in glucose and fructose deter-
        mination, 131
Molybdenum:
    determination:
        by catalytic current, 94
        by promotion, 78
    and induction, 80
    and selectivity in catalytic determina-
        tions, 44
Monitoring:
    amperometric, 93, 96
    in catalytic end-point indication, 75
Multicomponent determination, and Kalman
        filtering, 233
Multiparameter system, in COD determina-
        tion, 117
Multipoint methods, 218, 221–225
    amino acids determination, 118, 143
Mutarotase, immobilized, 104, 105

NADH, see Nicotinamide adenine dinucleo-
        tide
NADP, determination by chemilumines-
        cence, 163
NAD(P):FMN oxidoreductase, 164
    immobilized, 161
1-Naphthol, determination, 120
1-Naphthylamine, in selectivity of catalytic
        determinations, 44
1-Naphthylethylenediamine,
    in determination of sulfonamides, 118
    in nitrite determination, 116
Nephelometric determination, instrumenta-
        tion, 116
Neptunium, determination in mixtures by
        method of proportional equations,
        131
Nernst–Planck equation, 242
Nickel:
    determination in mixtures by differential
        rate method, 124
    in luminol chemiluminescence, 159

Nickel carbonyl, determination by chemiluminescence, 161
Nicotinamide adenine dinucleotide (NAD),
  54, 68
  determination:
    by chemiluminescence, 163
    in enzyme cycling, 69
    in fluorescence monitoring, 151
    recycling, 99
    of reduced form (NADH), 165
    spectrum, 65
    structure, 66
Nicotinamide adenine dinucleotide phosphate (NADP), 163
Ninhydrin, in determination of amino acids
  by uncatalyzed reaction rate method,
  118
Niobium, coating in graphite furnaces, for
  atomic absorption spectroscopy, 247
Nitrate, determination by chemiluminescence, 161
Nitric acid, determination by chemiluminescence, 161
Nitrilotriacetic acid, as activator, 78
Nitrite:
  determination:
    by catalytic currents, 94
    by chemiluminescence, 161
    in water samples by uncatalyzed reaction
    rate method, 116
  in sulfonamides determination, 118
p-Nitrobenzaldehyde, in PCC determination,
  119
Nitrogen oxides:
  determination by chemiluminescence, 161
  in determination of sulfur compounds, 162
p-Nitrophenyl esters, determination:
  by chemiluminescence, 166
  by differential rate method, 144, 146
Nitrosamines, determination by chemiluminescence, 161
Noise:
  in absorptiometric detection, 190–192
  flicker, shot, in fluorescence detection, 229
Normal-probability integral deformation, in
  air-segmented continuous-flow, 229
Nuclear fuels, see also Plutonium; Uranium
  determination of Tc in, 49
Nuclear magnetic resonance, 18
Nylon:

open tubular reactor, 106
  shavings, in enzyme reactor, 102

Open-tube reactors, with immobilized enzymes, 106
Operational amplifiers, 7, 171
  PIN diode/opamp circuit, 192
  in switching circuits, 204
Optical feedback, in light source stabilization, 191
Optimization:
  of catalytic methods, 44, 45
  of coupled chemical reactions, 222
Order of reaction, in classification of "slow"
  and "fast" reactions, 17
Organophosphorus compounds, determination in pesticides, 85, 88
Oscillating reactions, 80, 81
Osmium:
  catalytic determination, 37, 44
  in detection of gas or liquid leaks, 49
  determination in mixture with ruthenium,
  44, 131
  in Sandell–Kolthoff reaction, 258
Overvoltage, in catalytic currents, 91
Oxalate:
  determination by chemiluminescence, 163
  in differential rate determination of chloramines and chlorine, 143
Oxaloacetate, 67, 77
Oxidase(s), 55
Oxidizing agents, in catalytic determinations,
  38, 39
Oxygen, in chemiluminescence, 159
Oxygen electrode, in enzymatic recycling, 70
Oxytetracycline, determination in pharmaceuticals by uncatalyzed reaction rate
  method, 118
Ozone, in determinations by chemilumines-
  cence, 161

Packed-column enzyme reactors, 104, 163
Palladium, in heterogeneous catalysis, 88, 89
Paper, as support in phosphorimetry, 154
PCC, see 1-Piperidinocyclohexanecar-
  bonitrile
Penicillinase, immobilized on glass beads, 107
Penicillins, enzymatic determination, 64, 107
Perbromate, determination by uncatalyzed
  reaction rate method, 116

Periodate:
  in uncatalyzed determination of phenols,
    117
  use in determination of dyes in mixtures,
    131
Permanganate, in ethanol determination, 119
Peroxidase, horseradish, 67
  immobilization, 107
  inhibition, 107
Peroxides, determination by method of pro-
    portional equations, 131
Peroxo complexes, and catalysis, 40
Peroxybenzoic acid, in single-point method,
    4, 126
Peroxydisulfate, in oxidation of organic ma-
    terials, 115
Peroxyoxalate, chemiluminescence, in $H_2O_2$
    determination, 162
Pesticides, determination by enzyme inhibi-
    tion, 85
pH,
  effect on enzyme-catalyzed reactions, 61
  effect on rate coefficient fluctuations, 212
  sensing, 102
1,10-Phenanthroline:
  as activator, 77
  in apoenzyme preparation, 82
  inhibition of Os catalysis, 44
  iron(II) complex, and promotion, 78
  in uncatalyzed determination of organic
    species, 120
p-Phenetidine, in catalytic determina-
    tions, 2
Phenol(s),
  in activation, 77
  determination,
    by differential constant-current poten-
      tiometry, 197
    by single-point method, 127
    by uncatalyzed reaction rate method,
      117, 120
  2,6-disubstituted, 25
  enzymatic determination in water sam-
    ples, 65
Phenol red, determination in mixtures by
    method of proportional equations,
    131
Phenylalanine, determination by uncatalyzed
    reaction rate method, 118
p-Phenylenediamine, in determination of al-

dehydes, 259
o-Phenylenediamine, in determination of
    ascorbic acid, 153
Phenylhydrazine, determination, 120
Phosphate, determination in mixtures with
    silicate, 132
6-Phosphogluconate, in enzyme recycling, 69
Phosphomolybdenum blue reaction, in deter-
    mination of phosphorus, 116
Phosphomutases, 55
Phosphorescence, 18, 149, 153
  comparison with fluorescence, 153
  sensitized, 155
Phosphorimetry:
  extrapolation in, 155, 158
  holding media for room temperature, 154
  pulsed-source/time-resolved, 155, 157
  room temperature, 154
Phosphorus, determination by uncatalyzed
    reaction, 116
Photodiode:
  in absorptiometric detection, 192
  arrays, 192
Photoluminescence, 149
Photomultipliers:
  in absorptiometric detection, 192
  in fluorometric signal-to-noise ratio, 229
Photon emission:
  in chemi- and bioluminescence, 158
  in fluorescence and phosphorescence, 149
Phototubes, 192
pH-stat,
  determination of ketones, 202
  instrument, 8
  sensing, 68
Pin diodes, 192
1-Piperidinocyclohexanecarbonitrile (PCC)
    determination in illicit samples, 119
Pivampicillin, determination by method of
    proportional equations, 131
Plate height, in chromatography, 242
Plate theory, in chromatography, 240
Plug flow, in variable-time kinetic model, 256
Plutonium, determination in mixtures by
    method of proportional equations,
    131
Poisson distribution:
  in air-segmented continuous-flow, 249
  in chromatography, 241
Polarimetry, 189

Polarographic wave:
  and catalytic currents, 92
  in optimization of catalytic methods, 45
Poly (p-Amino styrene), in enzyme immobilization, 100
Polynuclear aromatics, determination by phosphorimetry, 155
Polyoxyethylene(20)-sorbitan trioleate, for chemiluminescence enhancement, 163
Polystyrene, in open tubular reactors, 106
Post-column detection, 47, 49
  by chemiluminescence, 165
Postkinetic currents, 91
Potentiometric methods:
  ammonia gas sensor, 117
  for detection, 102, 194, 197
    in continuous-flow, 256
  differential, constant-current, 197, 198
Power supply:
  regulation, 191
  stability, 192
Pre-equilibrium case, and catalytic mechanisms, 27, 55
Prekinetic currents, 91
Probability plot, in air-segmented continuous-flow, 251
Promotion, 72, 159, 259
  applications, 78
  effects, 24, 76
  and induced reactions, 80
Proportional equations, method of, 12, 128–130, 136, 137
  applications, 131
  critical evaluation, 138
  in time-resolved phosphorimetry, 157
Prosthetic group, and cofactors, 54
Protein, determination:
  by biuret reaction, 117
  by chemiluminescence, 160
Protein-bound iodine (PBI), 45
Protocatechuic acid, in catalytic determination of Mn, 49
Pseudo-first-order process:
  in atomic absorption spectrometry, 245
  concentration-time profiles, 213
  in differential rate methods, 124
  in error analysis, 215, 223
  in Kalman filtering, 231
  in uncatalyzed reactions, 111, 112
Pseudo-induction period, 64, 120

Pseudo-zero-order process:
  in catalyzed reactions, 26
  in differential reaction rate methods, 136
  in uncatalyzed reactions, 112
Pyridine, as activator, 45, 77
Pyridoxal 5'-phosphate, determination, 117
4-(2-Pyridylazo)resorcinol (PAR), in differential reaction rate methods, 124
Pyrocatechol, see 1,2-Benzenediol

Quantum yields, 159
Quenching, 130
  by dissolved oxygen, 153
Quinine, in determinations by method of proportional equations, 130
8-Quinolinol, see 8-Hydroxyquinoline

Radial diffusion, 238
Radial mass transfer, in air-segmented continuous-flow mixing, 180
Raman spectrometry:
  instrument for, 198
  scattering, 149
Random walk:
  in chromatography, 241
  in unsegmented continuous-flow, 253
Rare earth, determination, 131
Rate:
  of bromination (in acetaminophen determination), 119
  determination, in tangent method, 128
  determining step, 27, 28
  in fluorescence emission, 150
Rate coefficients, 17
  in ascorbic acid determination, 119
  in catalytic determinations, 26, 42
  in chromatography, 241
  comparison of precision in measuring, 230
  and error in uncatalyzed reaction rate determinations, 114
  in kojic acid determination, 119
  in method of Roberts and Regan, 135
  in minimization of errors in differential reaction rate methods, 224
  minimization of fluctuations, 212, 213
  in multipoint linear regression, 218
  and selectivity in catalytic methods, 43
  in series processes in continuous-flow systems, 255
  in single-point method, 127, 139

Rate coefficients (*Continued*)
  and stopped-flow mixing, 173
  in two-rate measurements, 218, 219
  in uncatalyzed determinations, 113, 114
Rayleigh scattering, 149
Reaction mechanisms, in error analysis, 212
Reaction plates, in packed-bed reactors, 104
Reciprocal time computers:
  digital, 201
  hybrid, 201
Recycling:
  of coenzymes, 99
  of enzyme solutions, 96
Redox reactions, in induction, 80
Reducing agents, in catalytic determinations,
    36–39
Regression analysis, 218, 220, 223–225
  nonlinear:
    in determination of amino acids, 118
    in differential reaction rate methods,
      142, 226
Relative standard deviation (RSD), in kinetic-
    based methods, 230
Resazurin, in fluorescence monitoring, 67,
    152
Residence time, in chromatography, 240
Resorufin, in fluorescence monitoring, 67,
    152
Retention time, in chromatography, 240
Rhenium, electrocatalytic determination, 93
Roberts and Regan method, 135, 136
  critical evaluation, 139
Roozeboom's triangle, and graphical inter-
    polation method, 146
Ruthenium, determination:
  in mixture with osmium, 37, 44, 131
  by oscillating reactions, 81

Sandell–Kolthoff reaction, 2, 45, 151, 258
Saponification rate, in single-point method,
    127
Saturation kinetics, in enzyme catalysis,
    57, 58
Sea buckthorn (*hippophae*) oil, determina-
    tion of natural antioxidants in, 120
Second-order kinetics:
  in differential rate methods, 133
  in linear extrapolation methods, 134
Selectivity:
  in catalytic determinations, 42, 43

  in enzyme activation, 81
  of enzymes, 54, 93
  in fluorescence determination, 151
Selenium, turbidimetric determination as
    selenite, 116
Sensitivity:
  in activation, 78
  in atomic absorption spectrometry, 246,
    247
  in catalytic determinations, 40
  in electrocatalysis, 92
  in fluorescence detection, 151
  in modified catalyzed reactions, 72
  in promotion, 78
  for protein determination by chemilumi-
    nescence, 160
  in series processes in unsegmented contin-
    uous-flow, 255
Sepharose:
  (4B), in determination of bile acids, 164
  (4B), in determination of several metabo-
    lites, 163
  as support for enzyme immobilization, 161
Series processes:
  in continuous-flow, 254, 255
  in nonflame atomic absorption spectros-
    copy, 245
  and transient signals, 96, 245, 254
Serum albumin, bovine and human, deter-
    mination by chemiluminescence, 166
Serum amine oxidase, in the determination of
    di- and polyamines, 166
Servo comparison, in measurement of slope
    of rate curves, 199
Side reactions, in initial rate measure-
    ments, 30
Signal conditioning, 14
Signal deformations, in air-segmented con-
    tinuous-flow, 249
Signal profiles, *see also* Transient signals
  for atomization of Cu, 246
  for penicillin determination with immo-
    bilized penicillinase, 108
  for sucrose and glucose determination in
    soft drinks, 105
Signal-to-noise ratio, in error analysis, 228
Silica framework, in enzyme immobilization,
    103
Silicate, determination in mixture with phos-
    phate, 132

Silica 'whiskers,' in enzyme immobilization, 107
Silver, determination:
  catalytic, 44, 78
  by inhibition of I⁻ catalysis, 46
Simplex optimization, of catalytic methods, 45
Simulation, computer, 76, 207, 253
Single-bead-string reactors, 107, 183, 254
Single-point method, 4, 12, 126, 127
  first-order (critical evaluation), 139
  first-order (plot), 126
  pseudo-zero-order, 136
  of Roberts and Regan, 135
  in second-order kinetics, 134
Single rate measurement, in error analysis, 212
Single species, determination by uncatalyzed reaction rates, 116, 118
Single switching network, 204
Slope, 198
  circuit for measurement, 199
  measurement, 29
Slower-reacting component, in uncatalyzed determinations, 113
Sodium, as scavenger in differential rate determinations, 137
Sodium thiobarbitone, as promoter, 78
Specificity:
  in catalytic methods, 42
  of enzymes, 54, 93
Spectrophotometric monitoring:
  of enzyme-catalyzed reactions, 63
  for kinetic methods, 189
$S_1$ State, lifetime, 150
Standard deviation:
  in chromatography, 241
  of mean dwell time, 250
Stat procedures, mixing in, 186
Steady-state process:
  approximation (condition), 28, 55,
  in fluorescence, 150
  signals, in flame emission spectroscopy, 248
Steepest descent, method of, 220, 227
Stern–Volmer equation, in method of proportional equations, 130
Steroid 21-hydroxy group, use in method of proportional equations, 132
Stopped-flow mixing, 172, 173

computer use in, 227
in differential continuous-flow methods, 140
fluid delivery system, 175
half-life in, 173
and immobilized enzymes, 107
by jet, 174
and method of proportional equations, 128, 131, 132
in multipoint methods, 224
and unsegmented continuous-flow, 176
Strontium, determination in mixtures with magnesium, 137
Sucrose, enzymatic determination, 64, 104
Sugars, determination in mixtures, by differential rate method, 137
Sulfanilamide, in nitrite determination, 116
Sulfate, differential determination, in atmospheric particulates, 141
Sulfide, determination:
  differential, in atmospheric particulates, 141
  by heterogeneous catalysis, 88
Sulfite, differential determination in atmospheric particulates, 141
Sulfonamides, determination in urine by uncatalyzed reaction rate method, 118
Sulfonephthalein dyes, determination by method of proportional equations, 131
5-Sulfo-8-quinolinol, see 8-Hydroxy-quinoline-5-sulfonic acid
Sulfur, determination in atmospheric particulates, 141
Sulfur compounds, 141, 142
  determination by chemiluminescence, 162
Sulfur dioxide (gas), determination by chemiluminescence, 162, 163
Switching networks, 204
Synergistic effects:
  in differential reaction rate methods, 138, 224
  in graphical extrapolation methods, 138
  and regression computations, 225
Syringe delivery system, 185

Tangents, method of, 31, 34
  in differential rate methods, 127
  in manganese determination, 116
  in perbromate determination, 116

Tanks-in-series model:
  in unsegmented continuous-flow, 254
  in variable-time kinetic model, 256
Tannins, and differential rate methods, 5
Tantalum, in furnace atomic absorption spectrometry, 246
Taylor expansion series:
  in differential methods, 227
  in Kalman filtering, 231
Technetium, determination, 49
Temperature, effect:
  in enzyme-catalyzed reactions, 61
  in rate coefficients, 217
  in stopped-flow mixing, 177
  in graphical extrapolation methods, 138
  in method of proportional equations, 129
  in selectivity in catalytic determinations, 46
Testosterone, determination by chemiluminescence, 167
Tetrachloro-$p$-benzoquinone, in determination of aliphatic amines, 118
Tetracycline, determination in pharamceuticals, 118
$\alpha$-$\beta$-$\gamma$-$\delta$-Tetraphenylporphinetrisulfonic acid, in chemiluminescence determination of albumin, 166
Thallium, determination by uncatalyzed reaction rate method, 116
Theory of propagation of errors, in minimizing fluctuations of rate coefficients, 215
Thermochemical detection, in enzyme-catalyzed reactions, 104
Thermometry, as approach to determination, 189
Thiamine, determination by uncatalyzed reaction rate method, 118
2-(2-Thiazolylazo)-5-dimethylaminophenol, in copper determination, 116
Thiocyanate:
  determination along with iodide, 140
  in heterogeneous catalysis, 88
Thiophene, determination by chemiluminescence, 162
Thiosalicylic acid, determination, 120
Thiosulfate, in heterogeneous catalysis, 88
Thiourea, determination, 120
Thiuram E, determination, 120
Thorium, determination:
  by ligand exchange, 117

by method of proportional equations, 131
  in uranium samples, 11
Thyroid hormones, determination with catalytic post-column detection, 47
Time-resolved determination, of sulfur compounds, 141
Titanium, promotion by, 78
Titration head, for continuous delivery of titrant, 185
$o$-Toluidine:
  in catalytic determinations, 49
  in uncatalyzed determination of organic species, 120
Toluidine blue, and heterogeneous catalysis, 89
Transaminases, 55
  glutamic oxaloacetic, 62, 67
  glutamic pyruvic, 62
Transfer coefficient, 244
Transglycosidases, 55
Transient signals:
  in atomic absorption spectrometry, 245
  in recirculation of enzyme solutions, 96, 98
  in unsegmented continuous-flow, 32
Transition state, in activation, 77
Transmethylases, 55
Transmittance, in catalytic determinations, 42
Transphosphorylases, 55
Triangular composition diagrams, in graphical interpolation methods, 146
Trimethylamine, in chemiluminescence determinations, 164
Trioctylamine, determination in waste water, 118
Triplet-to-singlet transition, in phosphorescence, 149
Triple-switching network, 206
Tripyridyls, in activation, 77
2,4,6-Tripyridyl-$s$-triazine, in uric acid determination by uncatalyzed reaction rate method, 119
Trypsin, activation by calcium ions, 81
$l$-Tryptophan:
  determination in food, 119
  kinetic fluorometric determination, 119, 153
  in pyridoxal 5'-phosphate determination, 117

Tungsten, determination in presence of Cr(VI) and Mo(VI), 44
Turbidimetric determination:
of lipoprotein, 119
of selenium, 116
Two-rate measurements, and error analysis, 216, 218, 219
Tyrosine, in enzyme immobilization, 103

Uncatalyzed reactions:
applications, 115
in error analysis, 227
rate methods, 111
Unsegmented-flow systems:
concentration profiles, 184
in differential reaction rate methods, 137
diffusion in, 252
manifold, 182
in solutions storage for stopped-flow mixing, 175, 177
Uranium,
determination, 45
by catalytic currents, 94
in mixtures by method of proportional equations, 131
Urea, enzymatic determination, 102, 103
Urease:
from jack beans, 53
reactor, 102
Uric acid, determination,
in blood serum and plasma, by chemiluminescence, 164
by chemiluminescence, 163
enzymatic, 64, 107
by uncatalyzed reaction rate method, in deproteinized serum, 119
Uricase, immobilized, operational stability, 101, 107

Vanadium:
in activation, 78
catalytic determination, 2, 12
determination:
with fluorescence monitoring, 152
in petroleum oils, 49, 117, 152

in induction, 80
Variable-time procedure:
error analysis, 227
hybrid computer for, 201
model of continuous-flow mixing, 256
and signal-to-noise ratio, 228
in sulfonamide determination, 118
switching network, 204, 205
in uncatalyzed determinations, 113
Variamine blue, in Tc determination, 49
Vibrational levels, in photoluminescence, 149
Viscosity, in mixing efficiency, 180
Vitamin $B_{12}$, see Cobalamine
Vitamin C, see Ascorbic acid
Voltammetry, catalytic currents, applications, 92, 94

Water:
analysis:
for bromine, 163
for $ClO_2(g)$ by chemiluminescence, 162
for manganese, 49
for nitrite, 116
for organics, 115, 118
for phenol, 65
determination in methanol by chemiluminescence, 160
hardness, determination by catalytic currents, 95

Xylenol orange, use in rare-earth determination by method of proportional equations, 131

Yterbium, determination in mixtures, 131

Zero-order process, in enzyme-catalyzed reactions, 59
Zinc, determination:
by enzyme reactivation, 82
in mixtures, 124, 125, 143
Zincon, in differential reaction rate method, 143
Zirconium, determination in mixture with Hf, 44

( *continued from front* )

Vol. 64. **Analytical Aspects of Environmental Chemistry.** Edited by David F. S. Natusch and Philip K. Hopke

Vol. 65. **The Interpretation of Analytical Chemical Data by the Use of Cluster Analysis.** By D. Luc Massart and Leonard Kaufman

Vol. 66. **Solid Phase Biochemistry: Analytical and Synthetic Aspects.** Edited by William H. Scouten

Vol. 67. **An Introduction to Photoelectron Spectroscopy.** By Pradip K. Ghosh

Vol. 68. **Room Temperature Phosphorimetry for Chemical Analysis.** By Tuan Vo-Dinh

Vol. 69. **Potentiometry and Potentiometric Titrations.** By E. P. Serjeant

Vol. 70. **Design and Application of Process Analyzer Systems.** By Paul E. Mix

Vol. 71. **Analysis of Organic and Biological Surfaces.** Edited by Patrick Echlin

Vol. 72. **Small Bore Liquid Chromatography Columns: Their Properties and Uses.** Edited by Raymond P. W. Scott

Vol. 73. **Modern Methods of Particle Size Analysis.** Edited by Howard G. Barth

Vol. 74. **Auger Electron Spectroscopy.** By Michael Thompson, M. D. Baker, Alec Christie, and J. F. Tyson

Vol. 75. **Spot Test Analysis: Clinical, Environmental, Forensic and Geochemical Applications.** By Ervin Jungreis

Vol. 76. **Receptor Modeling in Environmental Chemistry.** By Philip K. Hopke

Vol. 77. **Molecular Luminescence Spectroscopy—Parts I and II: Methods and Applications.** Edited by Stephen G. Schulman

Vol. 78. **Inorganic Chromatographic Analysis.** Edited by John C. MacDonald

Vol. 79. **Analytical Solution Calorimetry.** Edited by J. K. Grime

Vol. 80. **Selected Methods of Trace Metal Analysis: Biological and Environmental Samples.** By Jon C. VanLoon

Vol. 81. **The Analysis of Extraterrestrial Materials.** By Isidore Adler

Vol. 82. **Chemometrics.** By Muhammad A. Sharaf, Deborah L. Illman, and Bruce R. Kowalski

Vol. 83. **Fourier Transform Infrared Spectrometry.** By Peter R. Griffiths and James A. de Haseth

Vol. 84. **Trace Analysis: Spectroscopic Methods for Molecules.** Edited by Gary Christian and James B. Callis

Vol. 85. **Ultratrace Analysis of Pharmaceuticals and Other Compounds of Interest.** Edited by S. Ahuja

Vol. 86. **Secondary Ion Mass Spectrometry: Basic Concepts, Instrumental Aspects, Applications and Trends.** By A. Benninghoven, F. G. Rüdenauer, and H. W. Werner

Vol. 87. **Analytical Applications of Lasers.** Edited by Edward H. Piepmeier

Vol. 88. **Applied Geochemical Analysis.** by C. O. Ingamells and F. F. Pitard

Vol. 89. **Detectors for Liquid Chromatography.** Edited by Edward S. Yeung

Vol. 90. **Inductively Coupled Plasma Emission Spectroscopy: Part I: Methodology, Instrumentation, and Performance; Part II: Applications and Fundamentals.** Edited by J. M. Boumans

Vol. 91. **Applications of New Mass Spectrometry Techniques in Pesticide Chemistry.** Edited by Joseph Rosen

Vol. 92. **X-Ray Absorption: Principles, Applications, Techniques of EXAFS, SEXAFS, and XANES.** Edited by D. C. Konnigsberger

Vol. 93. **Quantitative Structure–Chromatographic Retention Relationships.** By Roman Kaliszan